Dear Penis (by David Allen Coe)

Dear penis,
I don't think I like you anymore,
You used to watch me shave,
Now all ya do is stare at the floor,
Oh dear penis,
I don't like you anymore.

It used to be you and me,
A paper towel and a dirty magazine,
That's all we needed to get by,
Now it seems things have changed,
And I think that you're the one to blame,
Dear penis, I don't like you anymore.

Penis sings...

Dear Rodney,
I don't think I like you anymore,
'Cos when you get to drinking,
You put me places I've never been before,
Dear Rodney, I don't like you anymore.

Why can't we just get a grip,
On our man to hand relationship,
Come to terms with truly how we feel,
If we put our heads together,
We'd just stay home forever,

Dear penis, I think I like you after all.

Oh and Rodney,
While you're shaving, shave my balls

This book is frank, informative and humourous – three important qualities to have when discussing the Johnson. I was really impressed with the topics covered too. The book had a nice flow to it while finding a way to include unique subjects about the penis that most people forget are important.

Greg Q. - Player

...you've taken a tough topic which is difficult to discuss, especially for guys who aren't good at talking about personal stuff and made it accessible. Most similar books are either too medical or too crass and I think you've struck a good middle ground. Nicely done.

Sag B. – Dad and Techie

That was a good read, and a good laugh. "Do you like the old man scraggly hair, ball sac look?" ...like how you bounce from accurate anatomy references, and the slang names all in a single paragraph. (testicles/balls, penis/wang). It make it sound real, and it doesn't bore you with scientific text only.
First word I can think of is "Frank".

Dave S. – Exec, Dad & Musician

Good coverage of a wide range of topics even if somewhat taboo. Funny and accessible, interesting, informative, quirky, useful, and fun.

C. B. Yoga mom & Techie

The tone of the writing was one of the things I liked most. The use of various penis lingo and other slang to do with this topic was refreshing. The book was professional where it needed to be and had the right amount of jokes to keep the mood light.

Greg Q. – Gov't worker, single

Good insight for women into how the bits work on a guy so makes it worth the read. The content is necessary for men and women in society. I learned from it - always a benefit.
V.S. Exec assistant & man watcher

Caution Please (important to read this):

This book is written with the understanding that the information shared with you is not diagnostic in nature, nor is anything in this book intended to be a prescription, recommendation, or cure for any specific mental, emotional, physical, sexual or spiritual problem. Should you be suffering from a sexual infection, testicle, prostate or other problem of a reproductive nature or other, obtain professional guidance. If you are experiencing any of these situations please make sure that you seek consultation with a qualified, licensed practitioner, therapist, or other qualified and competent professional.

Before starting any exercises presented in this book please make sure that your health is good and that you are cleared to have sex and/or cleared to exercise by a qualified health practitioner.

Thank you.

The Penis Protocol

A handbook to unlocking the mysteries of everything interesting, weird, wonderful, and wow about your weiner, willie, schlong, wanker or Johnson.

About the Author

Dr. James, a native to Ottawa, Canada, is a father, son, brother, chiropractor, author, and speaker. He has appeared on various radio shows, is a regular on "CTV Morning Live," and is the creator, writer, host, and producer of "MANtalks TV," a show dedicated to talking about sh*t that guys just don't talk about.

Over the past 30 years, Dr. James invested time and effort in order to study many health disciplines: from chiropractic, to acupuncture, homeopathy, yoga, craniosacral , energy medicine, native healing, spiritual work, meditation, Pilates, qigong, and other self-discovery work. He has implemented those disciplines into his personal life to make himself more clear, more true, and more whole.

The year that his separation and divorce occurred, his will was nearly broken. This was the compilation of many things as during that time, his dad had a stroke and needed ongoing care; his mom was diagnosed with cancer and died shortly thereafter; he was summoned to court as a witness; he had to deal with a crisis at his clinic; and he had to leave his matrimonial home to start his new life as a single father, forcing him to figure everything out on his own – to find out what he was really all about and who he really was.

At that lowest point in his journey, Dr. James made this promise to himself:

"I know I am going to survive this craziness, and on the other side of this chaos, I promise that I will find a way to make this (my situation) a template to help men. I promise I will serve men, to allow them to find for themselves certainty, confidence, authenticity, and a way to stand tall, spine straight, head up, facing the future no matter what it brings to them."

That promise was his driving force to write and create the "Guys with Guts" web series, which grew exponentially in viewership and morphed into "MANtalksTV", the show that talks about sh*t that guys don't talk about. That promise was his driving force to stand up and speak before rooms full of truckers and audiences of women, talking to them about men's health and well-being. That promise was his driving force to deliver this book to you.

Be Original
Be Authentic
Be Extraordinary
Every day

*No penises were harmed in the making of this book

Acknowledgements

To the many people who have played a role in this book and in my many projects to date I thank you.

Thank you to my brother, Jeff, my confidante and my straight-up guy. I tremendously value your perspectives, wisdom, and commitment to me on "how it is little brother." The four-legged stool stays with me for always. I love you.

To my Men of the Roundtable: Greg, Marc, and Michael. Thank you for your clarity, wisdom, fun-loving natures, complex simplicity, and the giving of your time in helping to make my vision become reality.

I would like to thank Suzanne M.G. for her inspiration with MANtalks and guidance with my other projects. I would also like to thank Venetia S., Veronica M. (who was the videographer for *Guys with Guts*), Charlotte B., and David S., who have contributed greatly to my projects.

Finally, I would like to thank my children, who, when they discovered what I was writing about, shrank with embarrassment into the seat of the car. Classic.

A memory forever...well, at least for me.

Dedication

Dedicated to my dad, a man of integrity, dry humour and authenticity. He passed shortly before this book was released. Dad was a man who possessed a deep unspoken and unconditional love, a love he shared with us through his actions. I am grateful for our life experience together.

This book is dedicated as well to all men, and the women and men they love.

Table of Contents

SECTION XIII - PENIS PROTOCOL CONCLUSION

Introduction

A simple thing, the penis...

It just hangs there, to the left or to the right; sometimes it stands up whether you want it to or not; sometimes it gets a nickname; it does need the occasional wash down, scratch, and blowjob. We intuitively develop knowledge that a kick in the balls hurts, jock strap cups are effective, wedgies are more fun to give than receive, and that Saturday night bar hopping searching for sex can actually be successful.

Some men know it is difficult to pee while having an erection, and that some guys' dicks are bigger or smaller than others. It also feels good when it is has pressure inside somewhere wet and warm. That is about a guy wants to know about his penis.

But there is more to the penis than that. From long or short members to multiple orgasms, from piercings to tattoos, from testicular exams to eating properly for the health of your testicles – there is a lot to consider and know.

This is not a book about sexual positions, nor a sex manual about techniques, although some of that is covered. It is a book about the awareness of your sexual nature, your power, and how to take care of your member from the inside out and the outside in.

At last count there were more than 200 names for the penis, and that, of course, does not include the name you have given your unit.

Regardless of the name you have chosen for your member, people around the world and throughout history have done some remarkable, forgettable, or infamous things with their penises. Pricasso in Australia, for example, creates paintings on canvas with his penis, for which he commands a high price tag. There are men who, through being guided or misguided in their efforts, painted their testicles and prostate areas on their perineum to gain increase sex drive. That testicle-painting thing, by the way, does not work.

There are museums for penises, with the most well known museum being the Icelandic Phallological Museum, which holds different phalluses from the tiny mouse penis to the gigantic whale penis. In fact, 56 different penises from 17 different sizes of whales can be found there.

Festivals celebrating the penis are held annually around the world, most notably Japan's Annual Penis Festival in Kawasaki – The Festival of the Steel Penis. This festival is all about a legend of a sharp-toothed demon that hid inside the vagina of a young woman and castrated two young men on their wedding night.

In order to rid the demon, they sought help from a blacksmith who made an iron penis. When inserted, the demon broke its teeth. Now the festival is held to raise money for HIV research.

The decision to write this book was the result of some significant personal and professional challenges that led me to transform my life. I discovered within myself that I wanted to make a difference in the lives of men in this world. As part of that promise, I would share what I felt would lead guys to experience greatness in their lives and ultimately lead men to live in their truth – to discover who they are and what they have to offer the world.

Those experiences and that discovery formed the basis of "Guys with Guts" and "MANtalks," and drove me to make a difference in the lives of men by looking at how I was being a man. It was interesting to discover, as I examined my life, how I had been conditioned by numerous internal and external sources to be a certain type of man in my marriage, a certain type of father, and a certain type of lover.

In viewing masculinity in today's society and, more specifically, exploring what men were all about, I knew I wanted to make a difference. Guys need more than what society is currently offering. I feel that guys are deeper and more multidimensional than the way society stereotypes men in this world.

Thus, I started writing – first blogs, then show scripts, then speeches, and now a book – about the basics of being a man...and a man's penis.

This is the Penis Protocol...enjoy the read.

Section I – The Beginning

1) Know the Anatomy & Physiology

Yes, you know that if you look down at your package, it has 3 parts – a penis and two testicles. No surprise.

But actually, the penis has five sections: the base; the shaft; the foreskin, which in some intact males covers the head of the penis; the glans (the head); and the meatus (the opening for semen and urine).

There are no bones in the penis, so the term 'boner' is actually quite inaccurate as the penis erection is made up of spongy tissue. Erections will be dealt with in deeper detail in a later chapter.

You likely already know that sperm is produced down low in the testicles, also called the balls or gonads, and is manufactured within their network called seminiferous tubules.

Along with sperm, the male sex hormone, Testosterone, is also produced by the testicles. This production can only be possible with the help of something called the pituitary gland, which is located deep within the middle of the skull, below the brain.

The penis and testicles do not work independently within the human frame. They function as a combination of connections to systems throughout the body, demonstrating that the body cannot be compartmentalized.

Whether you know it or not, everything in the body is connected. So much so that when changes in body function occur (for example, a change in the kidneys, pancreas (blood sugar) and/or liver (digestion and hormone secretion)), there is a dramatic effect on the blood levels of testosterone.

Have you ever wondered what testosterone really does?

Testosterone (or T) is responsible for many things of a sexual nature, including:

- Growth and development of prostate and male sex organs.
- Growth of male hair, such as leg, underarm, pubic, and facial.
- Deepening voice.
- Sex drive and sexual function.

When the kidneys are unhealthy, as in chronic kidney disease, the level of testosterone has been found to be lower. Diabetics have lower T levels as well, a result of insulin and other sugar hormones in the body.

Basic physiology shows that every male body has both testosterone and the female sex hormone, estrogen. This fact is important to know for the liver's influence on the body. When the liver becomes sluggish or dysfunctional, as in cirrhosis of the liver, there is a breakdown of testosterone, leaving estrogen to have a more dramatic effect on the male body.

When estrogen is elevated as a result of liver issues or other reasons, it can exert its feminizing effect on the male body. These effects can show up as a lack of interest in sex, shrinking testicles, and breast enlargement, which some guys would call "bitch tits" or "moobs."

If your testosterone levels drop below normal, your sex drive may lessen. You may also find it difficult to achieve and maintain an erection or grow body hair. In addition, you may experience a decrease in muscle mass. Low T can also cause osteoporosis, which is a lack of calcium in the bones, making them weak and brittle and more susceptible to breaking

But low T can also cause many symptoms unrelated to sex. Low testosterone levels can:

- Lower your energy levels.
- Cut your drive to get things done.
- Make you more irritable.

You may also find with low T that it's tougher to concentrate, while your risk of depression may rise. Since all of your body is connected, it is imperative for you to take care of all aspects of your body and its function.

The perspective of "if it ain't broke, don't fix it" is flawed. Burying your head in the sand and thinking your body is functioning perfectly every moment is an illusion. Every day your cells are shifting and are under pressure to change, a result of the toxins in the air, in the food and water you consume, and even in the negative thoughts in your mind.

Getting back to more anatomy, a very important set of muscles found "down under" are called the pubo-coccygeal (PC) muscles. These muscles aid in the control of ejaculation.

Check **Appendix A** on how to find the PC muscle.

If you look down when naked and can't see your package, then that is a problem. Find a mirror. If you haven't seen your penis or your feet in years, you are either losing your vision or your belly is big. Time to lose the belly. Stats show that the higher the ratio of waist to hips, the more likely you are to develop cardiovascular disease such as stroke, diabetes, or heart disease. Lose your belly and your heart, penis, and health will thank you. By the way, a penis looks noticeably smaller when the belly is larger.

Since your penis is a combination of all systems functioning properly, take care of your kidneys and digestive processes (stomach, liver, gallbladder, pancreas). Watch the food you eat, and take it easy on the jolly liquids. Also make sure to exercise, and keep a balanced mental outlook on life. It will all help to bump up your T.

Finally strengthen your PC muscles to aid in ejaculatory sensations. (See **Appendix A** for exercises)

Penis Protocol 1:

Work every day to take care of and know your anatomy.

2) Cut or Uncut – Not Talking Abs Here

We are talking circumcision. Guys, this is important to know because somewhere along the line you probably wondered about it, or why it's done. Not only that, but you may also face this decision if you choose to have children of your own someday.

By definition, circumcision is the surgical removal (through use of knife, rock, or scalpel) of the foreskin of the penis, that amount of skin that covers the head/tip of the penis. The foreskin has a complex array of nerve endings, between 20,000 and 80,000, resulting in numerous sensations. For those who choose circumcision, it is usually done on babies, but some men choose to have it done as adults.

Circumcision is recognized as the longest recorded planned surgical procedure of all time, most likely predating recorded history. The first known historical record for circumcision happened in Egypt as far back as 2400 B.C.

The prevailing thought is that circumcision began spreading from the Mediterranean with the Egyptians and Semites. It carried through to the Jews and Muslims, and then throughout Africa. Australian aborigines and Polynesians practice circumcision, and some evidence suggests that Mayans and Aztecs practiced it in the Americas as well.

Ancient Egyptians performed circumcision as a rite of passage from childhood to adulthood. This was a ceremony in which a sharp stone was used as the cutting tool. Many religions, including the Islamic faith, practice circumcision. In the Jewish faith, a bris is performed on the eighth day of life, a ceremony with traditions, rituals, and prayers.

Tribal groups have recorded and released songs, and have established incredible ceremonies around this practice of circumcision. For example, a tribe in Africa, the Abakwetha (this name means "group learning") teaches boys how to become men through circumcision. In the elaborate process, five youths at a time, from the ages of 17 to 20, are circumcised. They use a sharp knife with no medical anesthesia. After the circumcision, the group of five live together in a specially-constructed hut for three months while they undergo the healing and transformation from youth to manhood. For this prolonged ceremony, there are rituals of healing herbs, food preparation, ways of eating, body painting, costumes. It culminates with the ceremonial emergence from their hut. Upon that emergence, they are men.

The current practice in North America is considered a routine practice with little fanfare, with circumcision usually taking place within the first few days after birth. Many reasons are put forward as to why circumcision is/was carried out. In the early 1900s it was done to reduce the desire to masturbate. In the 1800s, prevailing thought was that it would cure paralysis, brass poisoning, and childhood fevers.

Moving to contemporary times, the roles of both world wars, the influence of a medical doctor, and the medicalization of childbirth had a profound effect on the rates of circumcision in the U.S. and the U.K. It reached a point with the military in the U.S. that they were requiring men to undergo circumcision to lessen the risk (or so it was thought) of contracting venereal disease. That mindset and a book from Dr. Benjamin Spock, who advocated circumcision, created a boom in the rate of circumcisions in the states. In the U.K., on the other hand, as a result of the ravages of war and the limited medical resources available for elective surgery, parents were not encouraged to circumcise their children. Subsequently, the rates plummeted in the U.K.

Today, people choose to circumcise their child because of:

a. Concerns over the attitudes from peers, and thus perhaps issues relating to self-confidence.
b. The ease of hygiene. It lowers the risk of HIV, STI, genital herpes, HPV, and syphilis.
c. The ease of circumcision as infants vs. the pain of an adult.
d. Medical benefits.
e. The father circumcised as an infant.
f. Religious or societal reasons.

Worldwide, approximately 30 percent of males 15 and older are circumcised, according to a 2007 report from the World Health Organization and UNAIDS.

Rates vary greatly depending upon religion and nationality, with almost all Jewish and Muslim males in the world having circumcised penises. Together, they account for almost 70 percent of all circumcised males globally.

Within the group supporting an intact penis, (i.e., no circumcision), there is a theory that circumcision creates a deformity of the penis. Some groups claim that by removing the foreskin, the holding elements of the penis are disrupted, forming scar tissue which can lead to tightness, discomfort, and even penile curvature or deformities. Those against circumcision also state that having the foreskin still intact is a blessing, not only for increased sensitivity for sex, but for proper penis growth, longer length, and personal hygiene. Proponents of the uncircumcised say that circumcision can cause post-traumatic stress disorder (PTSD) in infants and adults.

22

Those men who choose to keep their foreskin report a heightened degree of sensation, and an increased awareness of sexual pleasure. Not only personal pleasure, but pleasure for the partner. They state that the foreskin, when it rolls back, increases the girth, thus becoming more stimulating for the recipient. The glans, which is covered during the flaccid, non-erect state (and thus buffered from the rub of clothing), also becomes more sensitive to stimulation as the foreskin rolls back.

During erection, the penile shaft elongates, becoming about 50 percent longer. The foreskin moves back to cover the lengthened shaft, making it easier for the male to enter his partner.

In addition to its function in normal erection, the foreskin makes masturbation, now recognized as a completely normal activity, easier and much more pleasurable.

When comparing levels of sensation on the head of the penis in the circumcised vs. the uncircumcised state, the intact penises have the advantage. This advantage was seen in a study by Sorrels et al. (2007).* They studied 163 male subjects that included both circumcised and uncircumcised men. Nineteen points on the penis were tested using a touch-test sensor. The glans of the intact males was found to have a greater sensation than the glans of the non-intact males. The area of greatest sensation on the non-intact males was the **circumcision scar**. With the intact males, there were five areas on the foreskin with significantly greater sensation than the circumcision scar on the foreskin. Their conclusion: that circumcision removes the most sensitive part of the penis. *BJU Int'l vol 9(4), 864-9, Apr. 2007

Some men have actually decided to reverse circumcision and regain their foreskin that was removed. In order to regain the skin sheath around the glans, the penis skin is stretched with a weight so that it hoods the head of the penis once again. For those men who regain their foreskin with this process, not only do they claim that their sensitivity elevates, but they also prefer the look. It is healthier for them, and more natural to have foreskin.

Penis Protocol 2:
Be thankful for what you have and enjoy the sensations, whether you are intact or circumcised.

3) Cleaning Under the Hood

Do you value your penis? Of course you do.

The male body part, arguably, that men agree to value the most, is the penis. If this is the case, then men owe it to themselves to keep the penis healthy and in good shape, not only for today but for their old age. Knowing that your unit is hanging down there is not enough. Being aware that you 'should' wash your member is a good start, but there is more to it than that.

Here are some questions to consider:

How do you wash your privates? What is the length of pubic hair that works for you? Do you like the old man scraggly hair, ball sac look? Or an overflowing forest sprouting out of each side of your tighty whities? Are there strange smells coming from down there? Does your member have growths, scales, itchy spots, white blotches, or something else on it?

These are all things to consider and observe. If these things are there, figure out why they are present and what you can do about them. Please, do not just take it for granted that a white scaly spot showed up after using a condom in a new relationship. Figure out what is going on. There are so many things that could cause skin eruptions or skin flaking on the penis. There are some serious conditions, and others that are just an annoyance and within your control (some of these are covered in a chapter later on in the PP).

If you are experiencing sensitivity to something, find out if you have allergy to condoms, acidic saliva, the material in your underwear, or if you have a yeast infection spread to you from your partner or caused by your diet or choice of foods. Even if you do not smell anything unusual, clean it anyway.

Guys, you have to be in touch with your testicles and know what they feel like on a monthly basis. A testicle self-exam is definitely part of self care (see Appendix K). Having a wash down there after participating in a sporting event, workouts, workdays, or after sex is normal. Even taking a pee immediately after sex, condom or not, helps to eliminate any bacteria and reduces the chance of a UTI (urinary tract infection).

For those not familiar with self-care of the uncircumcised penis, here we go:

It is important to warm up the tissue of the foreskin, either during a bath or shower. Once the tissue is warmed up, retract the foreskin to clean gently where the foreskin covered the glans. Use only your hands to wash or scrub, as using a washcloth causes roughening and can damage your penile skin. Rinse thoroughly.

Lack of proper care under the foreskin can lead to build up of smegma, a white, cheesy, smelly substance. Not so great during romantic times when you are looking for some oral action. If you want to strike a sexy pose naked for your partner, the cheese curds piling up around your dick will most likely be a turn off.

What is the optimal cleaning frequency of care for those circumcised or not? Is there one standard practice or frequency? Who knows, as it is so individual, but regardless, soap is needed in either case. Avoid scented soaps as they can create irritation under the foreskin, and avoid soaps that dry out the skin.

Once clean, there are natural products called emollients that can be applied after you have patted yourself dry. Note that I wrote "patted," not "rubbed." Patting is easier on the skin.

These emollients help refresh the skin and can repair tissue damage caused by day to day living, incorrect choice of underwear, chemicals, harsh soaps, or micro-abrasions from washcloths.

See **Appendix C** for natural recipes of skin emollients that you can use right away.

Penis Protocol 3:
Wash thoroughly; dry properly, fully and regularly; then use natural skin conditioners on your whole body, including your package. Keep your package healthy and smelling pleasant.

4) Manscaping

Yes, manscaping!

This is the term used for the up-and-coming trend of attention men give to their body hair and body care.

Beauty spas are noticing a trend by men. Where 10 years ago the client list was probably less than 5 percent men, one spa owner told me that percentage has shot up to almost 30 percent. Spas for men are popping up and are specializing in hair removal from all areas of the body, including penis shaft hair, nose and ear hair, pubic and testicular hair, chest hair, butt hair, and butt crack hair.

Did you even know that your penis has noticeable hair up the shaft of it? Take a look.

While some men are having butt crack hair removed for aesthetic and pleasure reasons, other men are having it removed to help them with medical conditions. The one physical condition that some men have crack hair removed for is Irritable Bowel Syndrome. The hair near the anus is such an irritant for an already overly-irritated system, that wiping becomes very painful. Once the hair is removed, there is an increase in comfort level around the anus. With IBS, any degree of relief is welcome.

Another reason why men are getting manscaped is that their partner has asked them to get a "Bro-zilian." Just as there is no one type of woman who gets a Brazilian, there is no one type of guy that likes a Brozilian. You shouldn't need further description on that one.

If you are planning to do your own manscaping, "good on ya" as the Aussies say. But remember there are a number of steps, maybe as many as 10, to effectively prepare your lower area for the trim or shave. Although the hair removal is most likely not a walk in the park, preparing the skin properly for hair regrowth is just as important as prepping for the shave. The biggest drawback of any shave is the razor burn or ingrown hairs. Doing this prep work helps diminish the number of ingrown hairs and the discomfort, plus it gives it a clean, aesthetic look. Shaving properly with the proper products will eliminate razor burn.

Guys, you may be tempted to...but never, ever shave dry!

Doing the dry shave is a technique that will give you razor burn and predispose you to ingrown hairs wherever you choose to shave on your body. For manscaping at a spa, a specific type of wax is used to strip the hair off your body. You could probably do the stripping at home. However, these people have the wax, the talent, and the knowledge to do it properly.

As you already know, the skin of the scrotum/balls is not the same thickness, nor is it as sensitive, as your butt skin or chest, so a different type of wax is used (though the area still may be tender during and afterward). Men in particular have a more robust reaction to the hair wax removal (more screaming) when compared to the reaction of the women.

Some women (and men) love a completely hairless package, just as some men (and women) love a completely shaved pussy. Ultimately, it is about choice.

There are great natural products now available for men to use for a pre- and post-body/face shave. Witch hazel and aloe vera are some popular components of a post-shave lotion/salve, just to name a couple.

See appendix D for prepping your skin for a wax/shave

Penis Protocol 4:
Thoroughly prep your skin for any shave, and make sure you take care of your skin afterward in preparation for the regrowth, which will minimize the chance for ingrown hair.

5) Underwear– The Home for Your Package

Guys, throw out your "lucky underwear."

You know, the lucky underwear that's full of holes with a stretched waistband and stains. It offers no support and really has no sex appeal should you get lucky while wearing it. That underwear has to go in the garbage...now!

Furthermore, you have to know that the college days of wearing one pair of underwear for four days, (one day proper, one day backwards, one day inside out, one day backward inside out) are gone. You most likely make enough money now to buy more than one pair of underwear. It's probably best to have at least two pair extra, including one pair for every day of the week, unless of course you like to go commando. However, commando is not the most ideal form of hygiene for your package, nor for your pants or shorts. It is easier and cheaper to wash and get the smell and skid marks out of underwear than it is to get the smell and stain out of your tailored dress pants.

Here are some questions to ask about your underwear:

Do you stick to what you know, always wearing the same style? Are you trying to have a baby? Is your style the same style as what your dad has? Have you tried different styles? Or are you the no-holds-barred kind of guy who will take on new underwear fashion just because you want to try it?

There are briefs, also called "tighty whities." There are low-rise, mid-rise, high-rise and boxer briefs. There are boxers, low-rise boxers, trunks, jockstraps and thongs. The mix and match of styles, materials, and designers seems endless, in which one could get stuck in the seemingly endless choices quite easily.

See **Appendix E** for more detailed underwear style descriptions.

The choice of material for your underwear is very important.

Cotton underwear is okay to sleep in, or for light activity or exercise, but as the exercise becomes more intense and more sweating is involved, the cotton underwear get soaked and heavy. Fortunately, there have been large technological advances in material design for all types of clothing, underwear included. The biggest advances in dryness are a result of wicking technology. That could be wicking from polyester, nylon, or merino wool. Thankfully, this new wicking material moves your sweat away from the body and keeps you cooler, lighter in the crotch, and more dryness. Warm and moist environments can lead to fungus, which can lead to jock itch. It is easily treatable, but also easily preventable by making sure you are wearing breathable, clean clothing. Also, make sure you adequately wash and dry these areas regularly.

If you are trying to create babies with your partner, those testicles need room to hang and be cool. Tight underwear increases the temperature of your sperm and renders them less fertile.

When buying underwear, choose ones that:

- Are wicking from top band to bottom, to keep you cooler and dry.
- Keep constriction to a minimum (legs and balls).
- Do not ride up into the crotch.
- Do not chafe you.
- Are not cheap, as your balls deserve the best. Dollar store briefs will give you painful dollar store rashes and dollar store constriction.

- For workouts or an active lifestyle, find a comfortable, breathable fabric with good range of motion in its fibres in all movement directions
- Have a waistband of underwear that corresponds to your pant size.

You have to remember, **function, style** and **comfort** in your choice of underwear.

Some men like to wear boxers that are larger than necessary because it gives them more room to move. This may be important for sitting around on the couch, but when you are dressing to go out, larger than necessary boxers may actually be a detriment to your overall image.

When we look at men's style, it is important to control your image. To do that, keep your underwear concealed beneath your clothing. It is wise to choose your under-look in accordance with what you will be doing, and the social situation. For example, you do not want bulky winter woolens, or oversized boxers under a sleek suit, regardless of the cut length of the jacket. The thick underwear may bunch up and give you a weird profile, or asymmetrical butt cheeks.

Check what is being created now for men's underwear:

- Horizontal fly.
- Anti-odor technology.
- Utility pocket for phone and cash.
- Fabric designed to stay in place between you balls and your leg to keep both areas dry and cool.
- Low-rise underwear.
- Built-in access pocket over the front pouch for those intimate moments with yourself, or just to hold your balls when they are lonely.

- Anatomically-correct pouches.

Don't wear sexy underwear while working out. That is just weird. There is no support or absorptive component to that type of underwear. Please remember that buying smaller underwear to accentuate your frontal look will only serve to create discomfort for you.

Be smart with this; you don't need small-fitting underwear to show off your package.

Penis Protocol 5:

Control your image. Buy proper underwear – not just ones that you tolerate, but ones that are healthy for you, are breathable, fit properly, and give you the comfort, room and feel you want.

Section II – Your Manhood

6) Penis Shape and Size – Does it Matter?

Did you know that the world record for the longest penis, flaccid, is 13.5 inches? For most guys, that would be from your crotch almost to your knee. Compare that to the average Canadian penis that measures, on average, 5.5 inches.

The overwhelming response from women interviewed for this book was that size doesn't matter. There is also a lot of current research supporting that statement as well. Apparently, it is not the size, but what the artist does with the brush that matters. As one female friend put it, "There are talents that they have, that are accentuated." But all women stated that if the man has a small member, it is what they create, do with it, and how they harness its energy that either makes or breaks the woman's sexual satisfaction. However, most women did admit to the fact that girth is more desirable and satisfying.

On the contrary, there is a group of women, particularly those specializing in sex toys, that don't believe the small members can make great lovers rhetoric by saying, "I have never sold a 3-inch dildo. The bigger the better!" For those with smaller penises, keep in mind that the top five centimeters is the most sensitive part of the penis, and the first five centimeters of a vagina is also the most sensitive. Just saying.

With respect to penis size, the opposite problem is also true. The problem could be one in which the man is so massively endowed, that no matter what position, normal intercourse cannot take place. This can weigh mentally on the man as well. In reading interviews with the world-record-setting penis guy, he wrote about not being able to penetrate very far, for fear of hurting his partner(s). Instead, he made other forms of play the priority, with intercourse as the dessert so to speak. There are always ways around being too big or too small.

With that said, forget blowjobs for massive dicks, unless the woman is able to unhinge her jaw and create extra space. I spoke with one woman whose husband is massively endowed. She herself could not take it in her mouth at all, but her husbands' ex could actually fit the whole thing in from top to bottom. That woman, by her own admission, actually unhinged her jaw. In some ways, this type of jaw talent would be very much appreciated.

Men have tried all kinds of things to improve their penis situation (more on this in another chapter), including penis transplants. Since the first transplant surgery in 2006, that type of surgery has yet to catch on. There have been reported cases of successful surgeries, but in one case, the recipient was so adversely affected psychologically that he had it surgically reversed. In researching further into penis transplants, there was a documented case about one man who, as a boy, had his erect penis cut off after it was slammed in a door. Now as an adult, with just a stubby penis, he is able to have sex with his girlfriend, have orgasms, and ejaculate. He is strongly considering a penile transplant.

Then there was the case of Lorena Bobbitt who cut off her husband's penis with a knife. It was surgically reattached, he became fully functional, and he discovered a career as a porn star for a short period of time.

Penile deformities do happen as a result of circumcision, Peyronie's disease (a bent penis, sometimes making sex impossible), injuries during sex, or trauma in which during the healing process, collagen/scar tissue is laid down. This scar tissue, because of its inelastic nature, does not allow the penis to become straight when an erection occurs. The penis is actually bent to varying degrees, depending on the amount of scar tissue involved.

Guys, please remember that some curvature is fine and perfectly normal. Some penises can actually curve in the opposite direction (forward) as well. The curvatures become excessive in some cases, to the point that the penis is bent midway up the shaft at ninety degrees. That is when potential pain and injuries can occur.

To solve this problem, there are injectable drugs being tested which are showing promise for breaking down the collagen and straightening the excessive curvature. Surgery is another option for severe cases, which when done will "tighten" the other side of the penis to balance out the curve. However, this type of surgery runs the risk of post-surgical complications.

Men with extremely curved penises often tend to hurt their partners unintentionally during copulation. So the rule of thumb during sex is... if it hurts, change the position, otherwise trauma during the bed rocking may occur. The damage from a bent, short, deformed, painful penis, one that hurts during sex, might be incredibly hard on a man's self esteem and his physical being. So pay attention and know what your body is doing and feeling during sex.

Whether your penis is above or below the average length and girth; is the shape of a mushroom or button; is curved or ribbed; or has an anteater, beer can, or needle design, remember this:

Penis Protocol 6:
Love and appreciate what you have.

7) Can I Grow it Bigger?

If you have concern over the length of your penis, you are not alone. More and more men are thinking that their penis size does not measure up. Many men are finding or perceiving their member to be inadequate. This thought process has become so prevalent, that a new term has been coined: penile dysmorphophobia.

Whether it is the result of a lack of T in the last trimester of pregnancy when they were in the womb, or genetics, some guys just have small dicks. Some guys don't care, but others have considerable angst when they see themselves as not big enough. They want to look better naked. Some think that this anxiety over perceived lack of size came about as a result of viewing pornography. This pornography depicts men, with very long and large members, seemingly knowing how to use their dicks for ultimate satisfaction of themselves and their partner, something that the common man is striving accomplish.

A number of studies have been done which found that most men with penile dysmorph, when they were measured, were in the average size range. For whatever reason, they had overestimated the average penis size, thus underestimating their own. Penile insecurity and performance angst has created a billion-dollar industry in the line of penis products. Penis extenders, penile pumps, cock rings or penoscrotal rings, penis enhancers, multiple types of penis surgeries, and "Here, take this med and grow a massive dick" are all now available for you.

Since the penis has no muscle, and only spongy tissue, you cannot "work out" your penis like you can your triceps. As stated elsewhere in this book, approximately half of your "true

39

length" is buried inside your body in a muscle called the Pubo Coccygeus (PC) muscle. That is the muscle you want to get to if you are looking for strength and endurance (see Appendix A).

Vacuum devices are designed with a hollow tube placed over the penis with the one open end in contact with the tissue of the lower abdomen and above the scrotal sac. This tube has either an electric pump or hand pump to draw the air out and draw blood into the penis. The blood is then trapped with a cock ring, making intercourse possible.

Vacuum devices have to be used daily for those with erectile dysfunction (ED), and have not been found to have an effect clinically on the length of the penis. That's great for ED, but not for length.

Penis extenders are purported to work by applying constant traction along the length of the penis tissue. The device is attached just under the head of the penis, and the other end is placed against the "dink bone," or anatomically correct name of the pubic bone. When the traction is applied, the device extends the penis away from the body by pushing under the rim of the glans. The thought here is that the penis will adapt to the force of this traction and the body will lay down more cells, making the penis longer and thicker.

In order for men to get any sort of gain at all from a penile extender, the device must be worn on the penis from 4 to 12 hours a day for 6 months to see any results. Interestingly enough, however, both the International Journal of Impotence Research 2002 and the Journal of Sexual Medicine 2010 found different penile extenders that seemed to elongate the member effectively. A recent comparison of penal extenders (2014)* found the 4 best devices can increase penis length from 1 to 4 inches. However, no information was given for girth increases.

Penoscrotal rings have some promise of increased girth while the man wears it for sex. However, a penis turning blue and becoming cold and numb during sex is a distinct drawback. Requirements are that they can only stay on for 30 minutes. If you are some sort of sexual marathon guy, that amount of time is too short.

Lotions and pills have not been found to work at all. However, what the lotions may actually do is connect the man, emotionally and physically to his genitals through the act of touching, allowing him to understand his sex organs a little more. Subsequently, he may have more of a connection and groundedness to himself and, with that, develop more certainty of who he is and what he offers the world. With more certainty comes better performance.

Injections of fat to the penis have been attempted to gain size. The usual end result, however, was disfigurement or infection. At best, the fat cells were reabsorbed and the penis returned to its pre-injection size. Surgery to cut the ligament of the penis does actually lengthen the penis by 2cm, but only in its flaccid state. The length of the erection unfortunately does not change. However, with the ligament holding the penis being cut, the erection no longer is able to stand up; it falls over. So there is no real difference in the long run.

Eat well, maintain an exercise routine, and examine ways to boost your internal physiology, particularly blood circulation. Include in that body routine a way to balance your stress and mental outlook on life.

*Clinical Studies prove the effectiveness of Penis extenders: BJU vol.103(6) Mar. 2009; JSM6(2)2009.

Penis Protocol 7:

Whether big or small, all penises (for the most part) function the same way. Love what you have been given.

8) How to Measure Your Penis

This is the one where you may feel the need to cheat or add a little something extra to the measurement, or even round up if necessary.

You need a ruler, string, and measuring tape (the soft and flexible kind of measuring tape).

When you measure, it is only necessary to take the ruler or string to the tissue right above where the penis meets the body. First measure in the flaccid, non-erect position, then measure at full erection. You are measuring the length of the side that is on top when the penis is erect. Using a ruler and pushing into the body to "take the slack out" is not allowed in the measurement. Some also call that technique the "bone press," which will of course give you a greater length. Just have the ruler in contact with the skin. If the penis is noticeably curved, use the cloth measuring tape or the string to measure the length of the curve, then put the string on the ruler to check your length.

The girth, or circumference, is measured on the fully erect penis. Using the flexible, soft measuring tape, measure at the base where the erect penis comes in contact with the body.

Penis Protocol 8:
Measure what you measure, and your length is your length. It's a great conversation starter, but choose your **audience for that type of content.** Picking the wrong person to start that conversation with may get you a restraining order.

9) Staying Stiff and Sturdy

As young men, we had an abundance of frequent and hard erections. As men age, the frequency of erections diminishes and the hardness of the erections decreases. First of all, we have to understand how an erection is physiologically possible, and also remember that erections are a delicate balance of brain, blood, blood vessels, nerves, and hormones.

A flaccid, or soft, penis has arteries and blood running through it. In order to get the "woody," the arteries within your member have to widen and relax. When this happens, blood flows in and the arteries expand, filling up the paired set of spongy tissue chambers and hardening the penis. There is a six-fold increase in the amount of blood in the penis. The veins, which carry the blood away from the penis, are then compressed; the blood is thus "trapped." And with more blood flowing in than out, up goes the sail.

Yes, an erection is a hydraulic event, relying on fluid to increase the pressure in the penile tissue. Erections are also a chemical event. There is a chemical called nitric oxide that allows nerves to communicate with each other, and also with the arteries of the penis.

When the blood flow into the penis has been compromised in any fashion, the erection becomes weaker. Conditions that could weaken erections include cardiovascular conditions, diabetes, smoking, medications, prostate conditions, surgery, damage to the veins and/or arteries of the penis, alcoholism, recreational drug use/abuse, and mental issues such as depression, anxiety, guilt. Interference of the nerves from the spinal column to the pelvic area can also compromise it. Surprisingly, over-exercising can weaken your sexual response by lowering your testosterone.

This question usually arises out of self-discovery in the teenage years: Why can't I pee when I have an erection? Believe it or not, you can pee while holding an erection, but it is difficult. The reason you can do it is because the pressure of the swollen spongy tissue closes off the pathway to the bladder, so the sperm goes up and out instead of backwards and in, which would create irritation problems for the bladder. And when relaxed, urine goes down and out without being misdirected into the testicles where the sperm is formed.

The trouble, however, in peeing with an erection is that the urine goes up and everywhere, so do that in the shower or outside. If sitting on the toilet trying to pee while having an erection, you would have to bend very far forward, almost with your head on the floor, to get the angle you need to flow it into the toilet – a move that is manageable, but challenging.

So do you want stronger erections?

If the answer is "yes," you will want to take all the steps you can to strengthen those erections, and keep having those erections for as long as possible.

See **Appendix F** for the 10 steps to stronger erections.

Penis Protocol 9:
Eating properly, relaxing, and boosting your testosterone with proper exercise will provide you with the greatest opportunity to have strong erections.

Section III - PP Facts

10) Penis Protocol Facts

Penis length varies from 5-20 cm (2-8 inches).

The penis has 3 columns of erectile tissues.

The average length of a penis of someone 18 years or older is 5 inches.

Other names for "being erect" are tumescent and hard.

An orgasm is not necessarily the same as climax.

Testicles hang in the scrotum.

Testicles produce sperm and testosterone.

Testicles maintain a temp of 35 degrees C.

Testicles are made up of 100 yards of seminiferous tubules where the sperm is produced.

Sperm volume is 2ml or greater.

The pH of sperm is 7.2 – 8.

Male sperm carry the X chromosome.
Female sperm carry the Y chromosome.

Male sperm are lighter and are faster swimmers than the swimming female sperm.

Sperm speed is 1-3mm/min.
The speed of ejaculation is 28 m.p.h.

The prostate gland contributes to semen and helps with ejaculation.

Wet dreams occur when the penis is 70 percent of full erection strength.

Nocturnal erections happen during REM sleep and happen about 7 times a night in young men.

Morning erections are lost when T is low.

You may be able to pee while having an erection.

The foreskin is made up of mucous membranes similar to the inside of the mouth or inside the eyelids.

The earth could be re-populated to its current level using the number of sperm that could fit into an aspirin capsule.

Half the length of your penis is inside your body.

Minute quantities of more than 30 different substances have been identified in human semen.

Having three balls is called polyorchidism.

The average size of a dwarf/midget's penis is the same as that of a man of average height: 5.5 to 6.5 inches.

Penis Protocol 10:

Know the facts before making assumptions; some of them will surprise you.

Section IV – Talking About Sex

11) Ejaculation

Ejaculation is a normal part of life as a man.

Sperm, cum, chism, splooge, sauce, jizz, jissum, spunk, wad, blow a load, blast, bust some nut butter, plant a seed, pearl necklace, creamed, pop, ghost load, skeet, and vinegar strokes are all names for, or associated with, ejaculation.

From an early age, we have dealt with sperm by experiencing nocturnal emissions (wet dreams), premature ejaculation, masturbation, blow-jobs, participating in the creation of babies, playful and rough consensual sex, and as a natural part of the life cycle. Not surprisingly, sperm levels diminish with age.

With sexual stimulation, excited impulse signals are sent to the spinal cord, then up to the brain. The brain then sends chemical (or impulse) messages via the spinal cord back to the penis. This occurs throughout the sex interlude, then ejaculation occurs.

Being aware of our body is so important to our health and well-being. That includes knowing about ejaculation. There are phases of ejaculation: pre-ejaculation fluid (or seminal emission), followed by propulsive cum. The pre-ejaculation fluid is the little bit of wetness that you find in your underwear, or glistening at the tip of the penis, as you get more excited. As you get closer to cumming, signals are sent to the spinal cord, then the brain, and down again from the spinal cord to the muscles controlling contraction at the base of the penis.

When ejaculation is imminent, the testicles are pulled up close to the body. That is an important signal, as you can delay ejaculation if you know what to do at the moment the testicles are drawn up. This technique is described in the "lasting longer" chapter.

When the signal to ejaculate has reached its peak, the PC muscles start firing, as do those of the prostate gland. The vigorous contraction (happening every .8 seconds) that results shoots cum out of the penis. During the process of ejaculation, sperm (produced in the testes) passes through the ejaculatory ducts and mixes with fluids from the seminal vesicles, the prostate, and the bulbourethral glands to form the semen. Some guys' splooge shoots all over, far and wide, while other guys' just oozes out. Both are normal.

Something you experienced as a teenager is nocturnal emission or "wet dreams." That means ejaculation while sleeping at night. This is a completely normal process as the hormones are becoming turned up. This can start in the age range of 10 to 12. These wet dreams are very normal as an adolescent. However, if this nocturnal ejaculation happens as an adult, something else is going on. There may be a congested prostate gland, insufficient supply or weak nerves to the pelvic area, or perhaps is has to do with stresses of an emotional nature.

Sperm consists of amino acids, zinc, fructose, citric acid, lipids, glucose, calcium, magnesium, sodium, potassium, and protein, among other substances. Cum also contains a substance that decreases the acidity of a vagina, making it less hostile to your swimmers. When you are younger, you could pop out an erection and ejaculation numerous times a day, for many days in a row. As you get older, the frequency naturally decreases, but for optimal health over the age of 30, a frequency of 1 to 2 times a week is recommended.

When you look at your frequency of cumming, its colour, the comfort or pain level when you release it, and what the semen is composed of, you can be more aware of yourself and more aware earlier of potential health problems should they arise. Your splooge may naturally be white, creamy, or slightly grey. The colour of the sperm can change to yellow if you ingest large amounts of onions, garlic, or sulphur-containing food. The jizz may also be yellow if you have not had an ejaculation for a while. A yellow tinge is also a natural and progressive component of aging. A yellow colour or red tinge may come from a prostate infection or other type of infection.

Jizz has its own unique smell, but should not be foul. Pain with ejaculation is not normal. A consultation with a qualified health practitioner is recommended if pain or unusual characteristics of sperm are noticed.

A question came up (no pun intended) recently from a celiac. Is sperm gluten free? This is a first-world question, and the answer is that no one knows for sure. However, you can change the taste of your sperm by changing what you eat. Some people claim that smokers have sperm that is acidic or bitter; vegetarians' sperm is reported to be a neutral taste; meat eaters' sperm is said to be more pungent; and those who eat fruit are reported to have a sweeter taste. No one mentioned what happens when you eat asparagus or garlic.

Penis Protocol 11:
Ejaculation is a normal part in the sexual nature of being a man. Have sex at least "twice a week, twice a week."

12) Premature Ejaculation & Lasting Longer

PE stands for premature ejaculation, early volcano, dropping your bombs before the mission is completed, gun going off too soon, or the 100m dash. Being the most common sex problem for men, particularly in younger men, PE happens to 1 in 3 men.

There could be any number of reasons why PE happens. It could happen during your first time ever having sex. Or it could be the first time in a new relationship as you get to know a new body. Remember, it takes time to become familiar with each other's movements and rhythms. You could be overly excited and not recognizing the signals from your body, or your partner might do a new crazy, sexy move you weren't expecting that sends you over the line of no return. Other thoughts about why PE happens include: masturbating as a teenager and trying to get off quickly, thus setting up a neurological reflex pattern to shoot quickly when stimulated; not understanding how the body and the ejaculation reflex works; or penile oversensitivity to new sensations.

Other factors for PE could be related to the stress of rushing to ejaculate so that you won't be caught doing it. Sometimes feelings of guilt, unworthiness, or other forms of self-judgment about sexuality, or going against religious or parental teachings, can create the environment for PE to occur.

Given that the average time from insertion to ejaculation is approximately 3 minutes, make sure that your expectations for your performance are reasonable with respect to that average.

Being in touch with your body's feelings, functions and sensations is very, very important. If you are not aware of your body's sensations, the feelings growing in your body may be misinterpreted or not interpreted at all, causing climax to suddenly occur. This can also be the result of focusing too much on the partner you are with, trying to please your partner, and not checking in with yourself to see how your excitement levels are progressing. Feelings of body tightness, anxiety, or even over-thinking the situation can lead to PE.

The whole idea of a resonant relationship – of emotional or resonant intimacy – can play a large role in your comfort level during sex. If your relationship is in trouble, or has high levels of chaos or resentment, chances are you may rush to "get it over with." If you are expected to please or to perform, this also can lead to a lack of control and an early release.

There are also, of course, biological reasons for PE. Those include (but are not limited to): hormone level imbalances, abnormal reflex of the ejaculatory system, inherited traits, and inflammation of the prostate or urethra.

Do you want to last longer? One of the things that could happen during sex is getting so fatigued from too much effort that you cannot get it up any more. Orgasm then becomes a distant hope. So going forward, learn to pace yourself.

You need strong arms, shoulders, abs, and legs, along with a strong back, core, and butt, for dynamic sex. Just as important is flexibility in your back, legs, and hips. Flexibility is important to manoeuvre around the bed, or for that pleasurable new angle you accidentally found while in the kitchen or on the picnic table. Perhaps through effective training you can prevent SRIs (sex related injuries) from occurring.

Remember reading in the section on ejaculation about how just prior to cumming, there is a natural reflex that takes place in which the testicles are drawn up into the body? While engaged in any sex act, you can monitor your body for breathing and other types of cues, including position of the testicles. You could be going great and rocking the bed and realize that things are starting to progress toward climax sooner than you had hoped. This is the time to check your testicles. As your balls start to ascend, you or your partner can gently – and I mean gently – give them traction down. You cannot have ejaculation when the testicles are descended. This testicle-lowering technique is a great way to prolong your endurance.

A technique for PE that was developed by Masters and Johnson is called the squeeze technique. Close to the point of release, the man withdraws. He or his partner then takes the penis and applies pressure just below the glands on the front with the thumb, and the just below the glans on the other side of the penis (with the index finger on the glans above the ridge and the middle finger just below the ridge). Hold it to reduce further blood flow to the penis and until the feeling of cumming subsides.

Lasting longer is a result of a number of things. The above exercise above is a nice one. Check out **Appendix B** for the 9 things you can do to last longer during sex.

Penis Protocol 12:
Lasting longer is a combination of factors. Take the time to recognize your cues, understand your body, and what to do when ejaculation is imminent.

13) Do I Have to Cum?

Men grow up thinking that they have to cum during sex or the event just isn't complete. Your friends have orgasms when they have sex with their partners, or through masturbation. Porn stars shoot cum all over the place and regularly during filming. Men must have to make ejaculation part of the event, otherwise they are not successful at sex.

That is not true at all.

What if it is important not to cum?

There is something called Sexual Continence that is important for men to know about. Sexual Continence is the ability to engage in the act of sex without ejaculation. I can hear the chorus of you guys saying, "What? No getting off? But that is the best part!"

Knowing about sexual continence may provide another perspective on this long-held misbelieve, and may add another level to your sexual ability. There are many religious, spiritual, and healing practices that do not require, or even suggest, that the man ejaculate. For instance, in the Taoist religion, it is written by Emperor Tang in The Principles of Taoist wisdom: "If the man has a sexual relationship with a woman and his seed is preserved, his vital essence is preserved..." Vital essence, in this case, refers to his overall health.

Evidence such as from the Taoist quote above and from Traditional Chinese Medicine has shown that sex with ejaculation depletes your kidney essence, which both practices claim is detrimental to health in the long run.

Using that philosophy and expanding it further, ejaculation depletes you, whereas semen retention makes you stronger. In fact, having sex and not ejaculating is actually seen as very healthy spiritually and physically. If you are able to perform sex without ejaculation, it is said that you are mastering your sexuality.

As part of being that master of sexuality, you also become fully aware of your body and aware of the sensations within you and around you. This mastering of your sexuality is a mastering of the now, of the here and now, being fully present. Being present, you will develop intense mental focus and be able to channel your energy into your sexual encounter and other endeavours with strength and energy. You will even become healthier mentally, physically, and spiritually as well.

As you read this, it may become obvious then that having sex is not about getting off, but about the whole experience inside, outside, front and back, head, hands, heart, and soul. This is about complete resonance with your partner that is greater than the physical.

I can still hear the chorus of doubters... "Satisfying sex without cumming? You have to be kidding me!"

Sexual Continence is about being a gourmet chef of sex. Being able to withhold your semen, Sexual Continence is a show of willpower. Centuries have gone by in which this technique has been utilized with much success. There are practices from the Tantric philosophy, Daoist yoga, Yoga techniques that undoubtedly were derived from the Tantric, and the Karezza Method (see the complete outline included in **Appendix G**). The Karezza Method is a style and technique that allows you the freedom of intimacy without the need to ejaculate. It is truly an interesting read about sexual continence.

Tantric is a style of meditation and ritual, originating in India in or around the 1700 A.D. It allows you to find, then operate from, a different reference point – one not based in ego but in place called infinite consciousness. As it is described, once you learn Tantra, you begin to operate from a place of acceptance rather than a place of fear. In today's society, fear seems to be very prominent.

When people hear Tantra, they immediately think of Tantric sex. However, there is so much more to Tantra than sex. There is no way that Tantra could be covered completely in such a short space. Find a Tantric practitioner and take the full time to delve into it if you so choose.

Sex is only one small part of the whole practice of Tantra. In fact the tantric sex component is presented near the end of the sacred teachings. Once they, the students, have developed their spiritual understanding, the introduction to the importance of sex in this practice can then be addressed.

"Tantra is that Asian body of beliefs and practices which, working from the principle that the universe we experience is nothing other than the concrete manifestation of the divine energy of the godhead that creates and maintains that universe, seeks to ritually appropriate and channel that energy, within the human microcosm, in creative and emancipatory ways."
David Gordon White University of California
Department of Religious Studies

The quote above actually means that there is a defined interweaving of spirituality with the physical. The Hindu and Buddhist practitioners described it this way to dumb it down and to make sex practices tangible and understandable for the masses. The ultimate goal of Tantra, as it is written, is to be fully absorbed into spirituality and not needing anything.

The purists claim that the act of Tantric sex is neither a realistic nor an authentic practice if you have not undergone the spiritual path and teachings according to the true Tantric philosophy.

Having said that, you do not need to be a religious scholar, or authentically interested in Buddhism, Taoism, or any religion to start trying the Tantric Sex moves. You probably will not get all of the benefits as you would if you were fully immersed in the education of Tantra. However, you can benefit from knowing and understanding your body and the body of your partner better, along with discovering energies of each other that are new and perhaps exciting. All you need is a partner, the willingness to explore from both of you, then the space and time.

See the Appendix H Tantric Sex exercises to try out.

Penis Protocol 13:
There is no rule that says you have to ejaculate every time you have sex. Master your sexuality.
How many orgasms can your partner have in one session? How many can you have as well?
Learn and have fun.

14) Junk Sex vs. Gourmet Sex

When it comes to sex – vaginal, anal or oral, most guys want a combination of three things: heat, pressure, and moisture. Both junk and gourmet sex offer opportunities for such. However, there is a much different approach and meaning to the two. Research for this book revealed the following definition of junk sex. It was so all encompassing that not much could be added or taken away. Compliments to the Urban Dictionary:

"We all know what junk food is. And we know what happens to us when we make a steady diet of it. Junk sex is like junk food – not bad enough to avoid, but definitely not good enough to make a steady diet of. The effects of junk sex include outbreaks of unhealthy relationships and a malnourished emotional life, and self-destructive behavior like spending waaaay too much time at the gym. In addition to physical symptoms such as irritability, pain, and sexually-transmitted diseases. Junk sex, particularly media-induced junk sex, leads to a vicious cycle of empty sexual encounters and soul-sucking loneliness, and the obsessive preoccupation with our skin. No hard-and-fast definition (pardon the pun) can nail down the exact occurrence of junk sex, but it is very real and each person must define it for themselves."

In today's society, it seems there is not much time given to the process of foreplay and its importance to sex. It just becomes lick and stick. Too often there is a need for immediate gratification. People meet up and "you want sex, and I want sex. Ok, let's do it." Yes, that is easy, decreases the time involved to get to the point, you get off, your partner gets off hopefully as well, not much effort needed, and you are avoiding the no longer used word or process of seduction, in the long slow sense.

In the quick meet up scenario, you may be really great at sex, and have great sex as well. But with respect to conversation outside of sex and to the deeper meaning of sex making you grow and know yourself as a person...is it happening? Or are you taking the dopamine (the feel good hormone) hit that happens after sex and moving on to the next target?

In this junk sex way of intercourse, is it just a cooperative form of masturbation? Is it used as an escape from our current reality? Is it deadening our soul, diminishing the glow in our eyes, and then the act of sex just then becomes purely mechanical? There is an absence of deeper awareness of self and others when junk sex takes place.

Thank you to the Urban Dictionary for defining gourmet sex as "The art of fine sex." What is gourmet sex, though?

Gourmet sex is akin to planning a gourmet meal and knowing who you want to be there and exactly what you will serve and when. You know days in advance what you want to do. You plan the menu, what is in the first course, second course, and dessert. You go shopping, picking out the exact foods you want. What wine or drink will accompany each course. Then the planning of what you will do after dinner and dessert.

Gourmet sex is like the long slow seduction, almost romantic in nature. Does this mean you have to be in love with this person to have gourmet sex or romanticism? No. What it means is you are respecting yourself and respecting your sex partner by treating yourselves in a fashion of mutual appreciation and gratitude, as valuable and important human beings.

The fine art of gourmet sex is entirely up to you. Make it your creation, from you intimately for your partner, and ultimately for you. And this gourmet creation is exactly how the sex will go as well. Long and drawn out and intimately connected with

62

the desired result or an intense intimacy. Or it will be short and powerful with the desired result, which was a culmination of numerous discreet acts of intimacy (text, email, Snapchat, etc.), occurring over several hours or days. To cum or not to cum is entirely up to you. You are the master of your sexuality, the gourmand.

The junk food sex approach, which some people could call fine, would be akin to you noticing that your stomach is rumbling, so you get a frozen dinner out of the freezer, pop it in the microwave, push the buttons to cook it quickly, and then eat it just as fast while watching TV. When it is all done, you toss the leftovers into the garbage without looking and sit back down on the couch to finish watching the rest of the show. You get fed, but have no real plan. Lick and stick and done.

Both types of sex have their benefits and drawbacks. But what is happening here is how much you are respecting yourself and more importantly, how you are getting in touch with your own body. One of the things that may limit you in expanding sexually is how much you rely on habit, doing the same old thing because it is easy and you don't have to think much – get in and get it done quickly.

Body and self- awareness, with resonant quality sex, is key to the health and wellbeing in your life process at many levels. Could it be that gourmet sex perhaps offers the opportunity for a greater degree of self-awareness?

Ultimately, you want to know yourself, become self-aware, know and understand your body, and be able to be the best and healthiest that you can be. This will translate into all areas of your life, including profound intimacy and quality sexual performance.

In searching for guests to be on my TV show "MANtalks," I met and had a conversation with a young man who had quite an interesting perspective on relationships and sex.

At 23 years of age, he had decided to work on himself and develop himself to the best that he could be, inside and out. Well read and articulate, he regularly works out. That is not the surprising part. The surprising part is that he wanted to work on his physical appearance and do this internal work for growth so that when he meets the woman of his dreams, they will be on equal terms and he will be ready for her. How is that for perspective? Maturity? Planning?

This is the guy who would make gourmet sex a priority.

Penis Protocol 14:
Have sex as you'd like, understand what kind of sex suits you, and for what reasons. Just be true to that.

15) Becoming a Multi-Orgasmic Male

To become a multi-orgasmic lover, by definition, you have to be able to have 2 or more orgasms without a rest period in between.

Called dry orgasms, these orgasms occur without ejaculation. The benefit of dry orgasms is that you do not deplete your kidney strength or essence, and you retain and experience all of the possible physiological changes associated with sex. These changes would include accelerated heart rate and increased breathing without cum release. This, once again, has to do with being self-aware and knowing your body, knowing your build up, when to draw back, and levels of your body sensations.

History and literature are full of ways to become multi-orgasmic. This chapter will focus on the basics, drawing information from Tantra and from basic human anatomy.

Once again, the pubo-coccygeal (PC) muscle is important to know about. Recall that these muscles are the basis for sexual health and essential muscles to be able to become a multi-orgasmic lover.

To review, the PC muscle is a group of muscles in the pelvis, attaching the posterior aspect of a joint of the pelvis symphysis pubis (the dink bone, which is the boney part right above where your penis meets your body) to the anterior surface of the coccyx (the lowest portion of the spine; the bone you can feel right above your butt hole). These connections will be important to know about in a moment. The job of the PC is primarily to control the flow of urine.

(review **Appendix A** for location and exercises for the PC muscle)

65

Tantric sex is sex that brings into play the mind, body, and spirit. Tantric Sex is not necessarily about having prolonged sex, but about creating an environment in which the man can have dry orgasms and many of them. The average session of Tantric Sex can be from 20 minutes up to 9 hours long. With practice, and being with a willing partner, your path to orgasm then becomes about the mind, body, and spirit experience, not just the physical.

When the orgasm happens, it becomes a whole body experience, where the PC muscle, due to its attachments, has a rippling effect up the spine and into the grey matter of the brain. Thus, one can have a full-on whole body orgasm, particularly if one focuses attention and energy in the proper way to make it happen.

Some Tantric experts state that the best way to allow the whole body orgasm, but to stop the ejaculation when you feel it about to happen, is to do the following:

Immediately tighten you buttocks and your sacrum, tighten the low back muscles, contract the muscles between your shoulder blades, and drop the chin to your chest and clamp down on the jaw.

Penis Protocol 15:
Know your body, know your responses, know when you are building up to orgasm, and seal off the channel for ejaculation. You can do this many times in each session. No need to hurry.

16) ED - Erectile Dysfunction

"Good morning, good morning!" the ad sings as a man, seemingly pleased with his performance last night, dances up and down the street and into his office. This ad is appealing to men as stats seem to indicate that a large number of Canadian males over the age of 40 suffer from some form of erectile dysfunction, or ED.

The definition of ED is when a man cannot maintain an erection to satisfy his partner or himself.

Symptoms may include:

- Inability to get an erection.
- Inability to maintain erection after penetration.
- Inability to keep an erection long enough to complete intercourse.
- Erections not hard enough for penetration.

The most common cause of ED is damage to the arteries, smooth muscles, nerves, or fibrous tissue in or around the penis.

The causes of such damage are many, including, but not limited to: smoking, alcohol abuse, obesity, certain meds, cardiovascular conditions, diabetes, high blood pressure, kidney disease, low T, neurological disorders, prostate removal or disorders, stress, and emotional factors.

As you can imagine, there is a huge emotional and mental component to this condition of ED. This dysfunction may create in the man the desire to refrain from any intimacy at all, and that would include touching, kissing, fondling, or giving oral pleasure. Having ED diminishes certainty and confidence, leading some men into depression. Regardless of the cause, there are many therapies available for ED.

There are times when the nerves from the spine are not supplying proper energy to the body, and in the case of ED, from the spine going to the penis. Over the years I have worked on the spinal vertebra in a number of men suffering from ED. The reason for this problem potentially occurring is that the respective nerves (from spine to penis area) are trapped at the spinal level, and subsequently the proper, fully-expressed organ function cannot take place. Through chiropractic care, after a series of adjustments, the ED begins to be resolved (if that is the only contributing factor) and full function can be restored, allowing erections to return.

Other men have undergone acupuncture, homeopathy, naturopathy, Chinese herbs, counseling, Ayurvedic medicine, sex therapy, oral meds, penile suppositories, and surgery. Vacuum devices seem to work well with ED as covered in a Penis Protocol section elsewhere in this book.

Surgery may be attempted to improve the vascular flow, or to do an implant, but these are done as a last resort. Implants may be of two types, one that is a semi-rigid, malleable prosthesis, and the other is a pump device inserted behind the testicles, which is the hydraulic pressure pump with a switch. This little mechanism is squeezed to inflate the bags within the penis shaft with a viscous fluid. The fluid reservoir is located in the abdomen. When the sex session is complete, the switch is moved into the opposite direction. The penis bags empty and the penis deflates. Orgasm and ejaculation are possible with this type of device.

Penis Protocol 16:

Sex is important for a healthy body. If you are not getting it up or keeping it up, start with the basics, good food, relaxed state of mind, and water. Consulting a healthcare practitioner for further assistance is wise.

17) Becoming More Body Aware

In my 30 years of working in the fitness and health industry, the same theme continues to come up again and again: people are not being body aware.

In our top-heavy world, a world of immediate knowledge and over-intellectualizing everything, we have, in my opinion, lost the ability to see the profound miracle of the body and how it works. We offer a lot to each other every day, but if we can truly understand ourselves and how we work mentally, emotionally, physically, and spiritually, we can give that much more to the benefit of others and to ourselves.

Lack of body awareness is when you become numb to your surroundings, and/or not aware of you in your surroundings. You may not recognize or feel your feet in your shoes, or you may not sense the shirt on your back, the light wind on your arms, or the birds singing just above your head. With lack of body awareness, your senses become dulled, or deconditioned. A lot of times men are in a fog from a lack of emotional connection, improper food selection, or development of unhealthy habits.

When you lose the body awareness you become detached from your emotions, or written in a different way, you become armoured against the sensations of emotions. When you are not body aware, this armour serves as protection against those supposedly unwanted emotions that make us feel anything at all. It is the feeling that makes us human.

Those emotions are the ones that make us feel who we truly are deep down. For most people, that is a scary proposition, and as a result, hiding is easier than opening up to who you are and living fully engaged in life. Experiencing emotions is important, but being led by the emotions means you have lost control to them. When you lose control to the emotions, perhaps you are no longer living a fully engaged life.

Living a fully engaged life, full of breath, feeling, confidence, certainty, gratitude, and authenticity will lengthen your lifespan. The physiological responses in your body have been clearly documented in the literature that a healthy perspective creates a shift at the cell level to a healthier state.

Should you choose a self-restricted lifestyle, which is described as one in which you focus only on what you have to do, you are limiting yourself in the long run. When and if this happens, you mostly stop thinking progressive thoughts, devoid of self-expression and creativity. And in this inability to think to a deeper degree, in which you only just think to get by, there is a level of certainty and confidence that is lacking. The confidence and certainty that is lacking drags the man farther from self-awareness and shortens his life experience. Negative thought life and negative self-perception is extensive in society these days. Within 10 seconds of having a thought, the cellular physiology begins to shift either towards or away from health, depending on the quality of the thought.

You are body aware when you recognize foods that create mental fog and foods that benefit your mental well-being. You become body aware when you know which foods or people weigh you down, slow your enthusiasm or make you sleepy.
Being able to think clearly, from the depth of your experience, is the way to self-awareness and self-knowledge. One becomes body aware when you recognize the signals during sex that your orgasm is starting to build, and how you want to work with continuing the pleasure or experience your cum.

Too often we put up with situations or people that are not in our highest good because we don't know any different. We are scared of the outcome, can't think outside of the box we are in, or just do not recognize the situation as a detriment. An example of these situations may be an unhealthy relationship in which you just put up with it, thinking that is the way it is supposed to be. That is your lot in life, and there is nothing you can do, or you are scared to act.

Sometimes it is a duality in a marriage, long-term partner, or a string of one-night stands that you have to put yourself through because that is what is expected of a man. But deep down you are struggling with some type of unknown burden. If that internal conflict or mental fog is too strong and sex is involved, sometimes you just can't "get it up." Something is blocking your ability to have sex, and it is most times self-perception or an undiscovered lack of happiness with the partner or components of life that keeps the wind out of your sails.

Some men practice meditation or yoga for body awareness, while others do tai chi or qigong. Some men try hypnosis for body awareness or become vegetarians to make their body cleaner and mind clearer. In this effort, they are trying to get in touch with themselves. There are so many ways to get in touch with your body. Why not start?

Taking another angle at this, let's say the ability to be self-aware is based in your stomach and, in particular, what you eat. After all, digestion starts the process of food breakdown and body rebuild. What fuels you for excellence? Do you know why and how your stomach works the way it does? What happens to the gallbladder or your mental focus if you eat too much fatty food? Then if you do eat these things, how do you counter the body or mental reactions? Just because your stomach doesn't hurt when you eat, but you have a tremendous amount of gas, is that healthy? There are important little things that when added together can make or break a level of health and/or mental ability you are trying to attain. With an increased level of health can come an increased level of self-awareness, and vice versa.

Illness and debilitating disease can bring a profound level of self-awareness as well. Thorwald Defthlesen wrote a wonderful book titled "The Healing Power of Illness," which covers a lot of diseases that can affect us and their deeper meaning in life. Deepak Chopra talks about this, as does Louise Hay, and many others. This perspective on disease and its effects on living presents much information, all of which are fascinating ways for you to realize what is going on in life that , subsequently, allow you to become more self-aware.

I am not trying to turn you into doctors, but I am asking you to spend a little time figuring out how your body works. So when you do that, there is more awareness of your body. There is an easy exercise for body awareness that is not at all sex-related. It is designed to ground you and have you feeling your feet on the ground. See Appendix I.

You are body aware when you can take a deep breath intentionally to slow yourself down during your day. You are body aware when you feel your body starting to tighten during sex and you take the proper steps to lighten and loosen to keep the volcano simmering a while longer. This is about being a master of your body and awareness of your overall participation in life. Even more so during sex, you become so aware of your body that you can stop the ejaculation and experience only the orgasm.

When not body aware, it may be either you can't get it up or it is a lick-and-stick, getting to point of ejaculation and releasing without much thought, effort, foreplay, or emotion. Usually in these cases these men might live life the same way, in almost a robotic existence.

When you are having sex and you are body aware, your whole body is participating in the act. You feel your hair, you feel your knees and toes, your nipples, her wetness, stomach, skin and more. You are listening, smelling, feeling, sensing, knowing, and tasting the whole act of sex. These all melt together to become one overall sensation. This is what can make sex fantastic for you and your partner. This is gourmet sex.

Sometimes in our habits, patterns, or choices, we cannot break free and look at ourselves, our lives, our actions with fresh eyes. You perceive yourself as doing just fine and not in need of any guidance or direction. However, sometimes it is worthwhile to spice things up. Doing things outside of our box, or comfort level, usually requires encouragement or working through of some fear. Sometimes our imagination creates a limiting dynamic greater than the situation or fear ever was.

I am going to ask you that during the next time you have sex, that you become "one with the penis." This will create certainty, confidence, and control within you. All of your sensations, inspirations, breathing, creativity, perspirations, caresses, and movements will flow into your member and into creating a dynamic and fulfilling sexual encounter for you and your sexual partner.

When you are one with your penis during sex and you connect to yourself, all of this will go right into your shaft, and you and your partner will benefit. Listen and feel your body, get in touch with your body parts, listen and feel if they need to relax, slow down, and when you need to take a deep breath.

The society is so jacked up on immediate gratification that the way to self-awareness most likely will not be easy. But starting with little things, just getting back to the basics, will start the self-awareness process.

Penis Protocol 17:
Learn how to slow down. Find foods that give you mental clarity. Breathe, listen, and feel to that going on around you.

18) Thinking with Your Big Head

What are you thinking about during sex? Is there trouble getting up and staying up when a series of random distracting thoughts pass through your mind's eye?

One of the biggest things to help with that is going back to knowing your body. Knowing your body during sex can serve to keep you focused. Knowing your body allows you to experience all the sensations by focusing on them. If we don't know our body and a sensation comes up that we don't recognize, mentally we shut it down or cum rather than knowing and understanding where the sensation came from, acknowledging it, enjoying it, and making it all part of the experience.

Not living in your ideal, not eating healthy, or not living according to the way society, religion, parents, friends, or lovers want you to live can create internal confusion, conflict, or mental fog. These may have a direct effect on your relationship interactions and on your ability as a lover. Masters and Johnson discovered "sexual focusing" in the '70s which did not allow the mind to "loaf," as they put it. It is certainly old school, but does have a lot of value. See Appendix K for their material.

In having conversations with a lot of men about sex, some other interesting facts or practices that they are involved with have been brought forward. Many notables related to me, but here is one that you might find interesting:

Many men fantasize during sex that their partner is some other hottie – a movie star, or porn star perhaps, and that fantasy might get them feeling a certain way. Instead of modifying the partner, modify the perspective by transforming yourself instead. Transform yourself into the lover you know yourself to be, modeled after someone you admire, or for whom people all drool over. Imagine yourself to be that porn star who has all the moves and ability you wish you had. Visualize that person and become that person while you are having sex with your partner. Notice the change in you and the change in the responsiveness of your partner. She does not have to know. Just have fun with it.

Penis Protocol 18:
Mental clarity and outlook are important for getting and maintaining your sex ability.

19) Contraception for Men

Men have six traditional birth control options: abstinence, condoms, outercourse (dry hump, tit fuck, blow job, hand job), withdrawal, anal sex, and vasectomy.

Abstinence is 100 percent effective in reducing sexually-transmitted infections (STIs) and pregnancies. However, sex is healthy and necessary for proper function of organs and hormones of the body. Abstinence is probably very difficult to commit to in today's society. The semen is produced daily and is reabsorbed into the body when not expelled.

Condoms are for preventing pregnancy, transmission of sexually transmitted infections (STIs such as gonorrhea, syphilis, Chlamydia, and HIV), or both. It is used on the penis for vaginal, anal, and oral sex.

I once met a woman who could put a condom on without her hands. She used her mouth. You guys know how to put on a condom. It's always a good idea to do a quality check when you are in action to ensure that it's still on or hasn't broken mid-stroke somewhere during your session.

Condoms can be made of latex, polyurethane, lambskin, or polyisoprene. Lambskin condoms are made from the interstitial membrane of a lamb. However, lambskin condoms are porous and completely ineffective in protection against virus and STIs. They are effective for pregnancy control as the pores of the lambskin condom are too small for the sperm to pass through, but not small enough for viruses and STIs to be stopped.

Polyisoprene condoms are latex free and perfect for those allergic to latex. Other than that, they function exactly the same.

Not all condoms are effective when they are not worn properly. Make sure you are wearing it effectively and you know how to dispose of those used condoms.

Numerous researchers have undertaken the possibility of finding something medically related that may offer men a chance for contraception without the use of condoms.

There is something being developed in London, England, called the "Clean sheets pill." It is named that because it decreases or eliminates seminal emission but leaves the sensation of ejaculation and orgasm intact. Dr. Nnaemeka Amobi and his team at Oxford have described the pill as working to relax just the muscles that normally propel sperm-containing semen forward and out. Without the forward propulsion normally through the vas deferens, circular muscle contractions along the length of the penis essentially close down the passage.

This med is being designed to reduce or eliminate the emission of semen so men not only will prevent pregnancy, it also will decrease the spread of semen-born diseases, including HIV.

The hope is that this medication can be delivered via pills that men take before sex, much like Viagra.

RISUG : Stands for Reversible Inhibition of Sperm Under Guidance. The hope by researchers is to provide a cheaper, more reversible alternative to the surgical intervention of a vasectomy. With RISUG, a liquid polymer is injected into the sperm carrying tube, the vas deferens. It is designed to provide the man contraception that could last up to ten years. Basically, RISUG neutralizes the sperm, making them infertile. Human clinical trials of RISUG are starting in India.

Vasalgel: Inspired by RISUG, a similar polymer, dubbed Vasalgel, is under development in the United States. Rabbit research is now completed, with trials underway on baboons. Clinical trials on men were to start in 2014.

Ultrasound: Researchers in Chapel Hill are using ultrasound to heat the testes, producing a deep warmth within them. Remember from an earlier section that the testes hang low to keep their temperature low for fertility purposes. When this is used, the contraceptive effect of ultrasound exposure (depending on frequency used) has a duration of effectiveness from six weeks to permanent.

Gamendazole: As a side effect to cancer treatment, researchers in Kansas City found decreased male fertility. Although normal sperm amounts are produced, research indicates that the drug works by interrupting sperm maturation, making the sperm non-functional.

Adjudin: Also discovered by accident is the drug lonidamine. It causes sperm to be released when they are immature. However, lonidamine was much too toxic for the body, so a related compound, Adjudin, was created. It was created so that it is taken up only by the precise receptors in the testes where it is needed for contraception. It has to be delivered by injection; the results so far are short lived.

JQ1: Researchers in Boston and Waco have created JQ1, which blocks production of a protein in the testes is essential to sperm growth. Lab mice were given JQ1. Findings were that sperm were significantly diminished, as were their swimming ability. This, of course, made the mice infertile. Their sex drive remained unaltered, and after the drug was stopped, sperm production rapidly returned to normal.

Testosterone and Progestin If injected or absorbed through the skin, testosterone alters hormonal messaging and reduces production of sperm. When combined with a progestogen and rubbed on in gel form, a daily application has effectively suppressed sperm concentration in almost 90 percent of men, with few side effects. Current research is exploring the best combination of testosterone and progestin.

Both testosterone and progesterone have an effect on the sperm. Testosterone alters hormonal messaging. Combined in gel form, they suppress sperm concentration with few side effects.

Men can also have vasectomies. In this procedure, the vas deferens of each testes is cut and tied off to prevent the passage of sperm. Although not able to pass through, sperm is still produced and stored in the testicles, but they are neither used nor reabsorbed by the body. As a result, it can create an inflammatory process. Because of this inflammatory response, the immune system destroys them and you end up having anti-sperm antibodies. This causes a lower possibility if the vasectomy is reversed to becoming fertile again.

Guys, please know and remember that a vasectomy does not have a 100 percent guaranteed rate of success.

Penis Protocol 19:

Your call but contraception understandably does prevent a number of unexpected surprises.

Section V – Know the Score

20) Guys, Here are Some Sex Stats

For those guys who want the numbers to compare how they rank, how Canada ranks, here we go (information was taken from various Internet sources)...

Rankings of various countries General Sex behaviour:

Switzerland: 21 percent rate their sex lives as excellent
Spain: 25 percent rate their sex lives as excellent
Brazil: 84 percent have sex at least once a week
Netherlands: 22 percent rate their sex lives as excellent
Greece: has the most sex in the world at 164 times/year
 China: 78 percent have sex at least once a week

Largest Penis Size (global avg. 5.5 inches):

Congo Dem. Republic: 7.06
Ecuador: 7 inches
Ghana: 6.8
Columbia/Venezuela: 6.7
Lebanon: 6.62
Cameroon: 6.56
Iceland: 6.5
Sudan: 6.48
Jamaica: 6.42

Smallest Penis Sizes:

Korea: 3.7
India: 4
Thailand/Cambodia: 4
China: 4.3
Ireland: 5
USA: 5.1

Canada is exactly at global average of 5.5"

Who lasts longer in bed (intercourse only):

Australia: 4:02 minutes
US: 3:45
Canada: 3:41
Russia: 3:31
France: 2:53
Italy: 2:50
China: 2:40

Penis Protocol 20:
Now you know where you rank; do with it what you'd like.

Section VI – The Prostate

21) The Prostate

Remember from the "Know Your Anatomy" section that the prostate is that little gland that assists in forming sperm, located in the perineum, between your anus and penis.

The prostate gland is part of the urinary and reproductive systems of the male and requires T to function properly. There is evidence to suggest that women also have a prostate gland, called Skene's gland. Both create measurable **prostate specific antigen** (PSA) when the gland has physiological changes within it.

The prostate is just below the bladder. Its size can be anywhere from that of a walnut to a small apple. Its 2 semicircular lobes (left and right) encircle the urethra, which is the tube that carries urine from the bladder and down through the penis.

The PG has two functions. One function makes use of the PG's muscular fibres and helps control the flow of urine, as it is able to squeeze the urethra that carries urine from the bladder to the outside. The second function involves the PG secreting its fluid. When ejaculation occurs, approximately 80 percent of the semen is made up of fluid from the prostate. This prostatic fluid drains into the urethra from the tiny fluid-producing glands interspersed in the network of vessels and muscular fibres of the prostate to join in the fluid produced by the testicles.

When looking at men over 50, the numbers appear to indicate that up to 50 percent of North American men have an enlarged prostate. By age 80, the numbers escalate to 80 percent of men having an enlarged prostate.

Why the diagnosis of enlarged prostate becomes important is that if left untreated, the urine flow could be diminished or blocked, leading to backed up kidneys and chronic kidney disease.

Dysfunction of the PG can lead to several health problems:

a.   Prostate enlargement, called benign prostatic hypertrophy.
b.   Inflammation of the prostate or prostatitis.
c.   Prostate CA is the second leading cause of death in men after lung cancer. One in 8 men will be diagnosed with prostate cancer in their lifetime.

Penis Protocol 21:
Massaging your prostate gland helps to keep it healthy. You might have a new type of orgasm happen as well.

22) Bicycle Seats and the Prostate

You may recall the remarkable story of bicyclist Lance Armstrong's recovery from prostate cancer and how some theorized there was a link between bicycle riding and the development of that disease.

Research seems to show that there is no direct or indirect link between cycling and prostate cancer, or prostate enlargement. There is weak evidence that trauma from bicycle riding can irritate the prostate and could exacerbate it, perhaps leading to prostatitis (inflammation of the prostate) or chronic pelvic pain syndrome. Only a small number of riders have been found suffering from prostatitis.

Although the seat, or saddle as it is sometimes called, is not linked to prostate cancer, it can irritate a prostate that is inflamed by other causes. It is the narrow angle to the tip of the seat that sits right in between the anus and the testicles, exactly where the prostate is located. Pressure on the perineum, that area between the anus and penis may cause numbness from compression of the Pudendal nerve, incontinence, and temporary erectile dysfunction. Increasing severe symptoms may be found in older cyclists with benign prostate enlargement.

Remember from an earlier section that PSA levels have been used in the past to determine health of the prostate. Well, it is known that cycling may temporarily increase a man's PSA level. In experimental studies, cycling causes an average 9.5 percent increase in PSA in healthy male cyclists greater than 50 years old when measured within 5 minutes post-cycling.

If you find it necessary to go for a PSA test, keep in mind the history of false positives and false negatives that have occurred with this test. To put the odds in your favour of getting an accurate number, remember to avoid cycling and sex with ejaculation 24 to 48 hours prior to the test.

The main problem with cycling then is how to reduce that pressure on the perineum and prostate gland. This can be done in a number of ways: wearing padded shorts, regularly standing on the pedals, considering the adjustment and position of the saddle, and actual saddle choice.

Therefore, one has to find a saddle that is designed to reduce pressure on the perineum. This search is important because with a better seat, there is less of a chance of it causing penile hypoxia, perineal numbness, or erectile dysfunction. Prostate, groin and perineum-friendly saddles, which have been designed to reduce pressure in those areas through the following:

- Grooves and holes cut out.
- Holes cut out and a cutaway at the back.
- "Split saddles" which have two sections and no central area.
- Noseless saddles.

Saddles with holes cut out of them may increase pressure inadvertently or pinch the perineum. As there are so many numerous designs, there is no correct answer, and the most expensive saddle is not the best.

Those who want to return to riding, and painless riding, have to switch to a new seat designed to relieve pressure in those prostate-sensitive areas. Any time a rider switches seats, the body position on the bike has to be modified as well.

Measurements must be redone to fit the rider on the bike to ensure efficiency, and new riding habits – handle, foot, and leg adjustments – must be made to ensure fluidity on the bike.

Using a wide rear saddle, the following has been found, courtesy of the ACSM (American College of Sports Medicine: "Bicycle seats with a wider rear saddle area for support of the "sit bones" and a narrow nose for balance and bike control put pressure on the crotch as well as the bottom of the pelvis. Incorrect adjustment worsens the problem. Tipping the seat up puts more pressure on the perineum, while tipping the seat too far down slides the body forward onto the nose of the saddle. Placing the seat too far forward or backward on the rails also causes pressure problems. Nosed saddles with cutouts for the perineum only partly ease the trouble. Adding padding doesn't solve this problem either."

If you have prostate problems, the best type of bicycle seat should support the bottom of your pelvic bones, but place no pressure on the perineum. The no-nose seat has either two separate pads or a single-piece saddle, but no part of the seat extends between the thighs. The National Institute for Occupational Safety and Health discovered the sexual health benefits of the no-nose saddles when studying the riding habits of bicycle patrolmen.

Penis Protocol 22:

Aerobic exercise is a great way to have a positive effect on the prostate gland. Whether it is on your street bicycle or the bike at the gym, attempt to diminish the pressure on your prostate gland. Aerobic activity without prostate compression has been found beneficial.

23) The Male P-Spot

You know where you liked to be licked, touched, and kissed. Those areas would most likely be your erogenous zones. By definition, an erogenous zone (EZ) is that area on your body that, when stimulated, may create sexual fantasy, sexual arousal, and orgasm. Yeah, there are the standard EZ spots – groin, glans, inner thigh, chest, ears, bag, neck, etc. We have all had them explored. However, the one male EZ that is truly the kicker for a potent orgasm is found inside you. This spot is found inside your rectum.

The point you are looking for is the approximately the size of a walnut, and approximately 2-3 inches on the anterior wall of the rectum. When you find this spot, it is actually the prostate gland that you are looking for. The prostate gland helps with seminal fluid, and is usually associated with the doctor-finger-in-butt-check once a year. In this case, the prostate gland is a source of pleasure, not for something to worry about.

Look guys, this P-spot (the guy's g-spot) is inside your rectum and you don't have to be gay to find it or work it. This P-spot is actually right on your prostate gland. Maybe from social stereotyping, religious upbringing, or something else, some men shun this type of intimate sexual experience. As a result, this type of sexual play is thought to be for gay men only, though working this spot won't make hetero men homo, nor will it make homo men hetero; it is just a spot for pleasure to boost your orgasm experience.

In order to find this spot inside, you will need to:

Trim your nails.
Get some natural lubricant, and a lot of it.
Use rubber gloves (highly recommended).

The rest of the instructions are found in **Appendix B**.

One can have an anal orgasm or stimulate the P-spot, and at the same time your partner is stimulating your penis, for an intense orgasm. If you are going to be doing that, why not maximize your pleasure?

Penis Protocol 23:
Maximize your sexual pleasure by knowing where your EZs are, plus understanding if P-spot massage is for you.

Section VII – Food for Your Package

24) The Erection Diet

I once gave a talk to a room full of hardcore truckers. There were the young enthusiastic guys new to the job, and the lifers who looked exactly as you might picture – ball cap, plaid work shirts, and big bellies. Great guys. They were a hilarious bunch, and challenged me at every turn. For good measure, they hooted and hollered for the things they liked. It was a test for them and for me at finding out what a traditional blue-collar man was all about. I learned a lot from them that day, and had a blast with them as well.

Two of the things they liked had to do with sex. The first one was the idea that having sex 2 times a week would increase their lives by 3 to 8 years. That went over fabulously well. So well, in fact, that I got an email two months later from the woman who brought me in to do the talk that the daily chant at the trucker dispatch after getting their driving assignments was "twice a week, twice a week."

The other one they hooted about was the food list I gave them. I called it the Rocket Pack, or in other circles, the Erection Diet. They were to eat it for lunch every day for one month, weekends off if they chose, and see what happened.

All foods were selected to stimulate circulation to the nether regions, and give energy to the body. The secret ingredient in this is arugula. In the Mideast, arugula is called Rocket fuel because it gets the rocket fired up. In **Appendix C,** you will find the Rocket Pack recipe. Try it for a month and let me know how it goes.

Penis Protocol 24:
Eating well helps your erections. Eat foods that stimulate circulation and foods that boost your testosterone.

25) Eating for a Healthy Penis and Testicles

If you remember anything at all from this section, it is this guys…. **Zinc** for the dink.

Zinc is vital to maintaining the health of cardiovascular cells and the thin layer of cells that line blood vessels called endothelium. This endothelium plays a major role in circulation, and since your woody relies on blood vessels, as do cardiovascular functions, their health is directly tied to zinc levels. High cholesterol buildup and inflammation result from low zinc levels, which create a weakness in the endothelial layer. Cholesterol buildup and inflammation increase your risk of cardiovascular disease.

Where would you find zinc? Zinc is found in high concentrations in oysters, veal liver, and roast beef. Various nuts and seeds have high levels of zinc, including watermelon seeds and pumpkin seeds.

Magnesium is the other one that is an essential contributor to the testicles making T. Mg is found in tuna and halibut, bananas and figs, and milk. Many grains, nuts, and vegetables contain magnesium, including most leafy green vegetables.

Interestingly enough, without proper levels of magnesium in the body, your lean muscle mass decreases, and guess what? With decreased muscle mass your T levels drop. So then you would have less muscle mass, less capacity to do exercises, and less endurance for activities such as sex.

Sexual arousal basically begins with testosterone acting on the brain. With low T, less arousal takes place.

Guys, you also want to start eating foods from the "testicle tree." Avocados. The ancient Aztecs referred to avocados as being grown and picked from "testicle trees." Avocados and leafy greens, such as spinach and kale, plus citrus fruits, and beans, are all sources of **folate** that you need for healthy sperm. Men low on folate may have 20 percent more unhealthy sperm (swimmers that have missing or extra chromosomes) than those with higher levels. Folate is a B-vitamin necessary for cell production.

Testicles produce testosterone. Men have to eat these foods that benefit the testicles for the production of T. It is the Leydig cells of the testicles that secrete the testosterone.

If you have low T levels, selecting foods high in certain fats, vitamins, and minerals will bring your low T up to normal. No amount of specific food will overdrive it. So men, you have to ensure you eat the right foods so that the testicles can secrete the adequate amount. With the proper diet, you will keep your lean muscle mass, lose excess fat, and bring your hormonal function into a balanced range.

To add to this information, here is something you can watch out for yourself about low T. An interesting symptom for low T is the loss of morning erections. When T is low, morning erections disappear. In order to restore the proper levels, T supplementation is taken through diet modifications or other sources, and subsequently morning erections return.

These 'nocturnal' morning erections are completely normal, are testosterone driven, and have little to do with erotic dreams.

Penis Protocol 25:
Start eating those greens, fish, and nuts

26) Breathing for Your Penis and Testicles

Put one hand on your belly with thumb around your belly button and other hand on your upper chest. Take a breath in. More often than not, the belly hand does not move much and the chest hand is pushed up. This is opposite of what needs to happen for proper sequence of breathing for optimal health.

As adults, we have forgotten to breathe properly. If you watch little kids up to the age of approximately seven, they breathe through their belly. When breathing in the belly expands, breathing out the belly retracts. This is normal diaphragmatic breathing.

As adults, we can get to a calmer place and more balance internally if we do this diaphragmatic breathing daily, especially during periods of stress.

If we are able to transform breathing, we can transform our health and physiology. Start with breathing 5 breaths like that before turning the light off for the night. As you become better at it, work up to higher numbers of in-breaths, and then selectivity use this technique to calm yourself down where necessary.

Having worked on balancing your breathing to improve your physiology, it is time to breathe into your balls. Breathing into your balls, is a way to ground you in your sexual energy, and bring you into connection with your sexuality.

Sometimes people can get energy stuck in their head, throat, and chest. This could come from thinking too much, not speaking our truth, holding it in. This can influence the way that energy flows into and around the body.

Breathing properly and into your balls opens up these "stuck" energy areas and allows your body to flow more evenly and have more sexual flow.

See Appendix M for the details on how to do this breathing-into-balls exercise.

Penis Protocol 26:
Work on your breathing, not only to de-stress, but also to ground yourself in your sexual power.

27) Eating for a Healthy Prostate

Remember in the earlier sections there is a description of location and function of the prostate.

When the prostate is not working well and is swollen, the symptoms may include: diminished urine flow, difficulty starting or stopping urine flow, dribbling after finishing, having to get up at night to pee, and/or pain in the lower abdomen.

Fifty percent of men over the age of 60 suffer from an enlarged prostate or benign prostatic hyperplasia (BPH), according to the Mayo Clinic. By the age of 85, more than 95 percent of men will live with BPH.

Benign, when speaking of a tumor, refers to the fact that the hyperplasia has not spread into surrounding tissues. Thus, it is non-cancerous. Hyperplasia refers to the increased rate development of cells within a tissue, usually indicating a cancerous situation. So BPH means that within the prostate there is a swelling of that gland due to overproduction of cells, but it is non-cancerous.

Risk factors for BPH include cholesterol, family history, aging, high waist to hip ratio, and ethnicity. The good news is that a diet rich in certain vitamins and minerals can keep your prostate healthy and lower your risk of BPH. And because being overweight is another risk factor for BPH, making healthy food choices is a great way to lower both your weight and your risk.

The health of your prostate is dependent on 5 factors. Three of those factors are not modifiable: age, ethnicity, and genetics. The other two, both modifiable, are food and lifestyle.

As your age passes 50, the risk for prostate cancer begins to increase. However, men above 65 is where the prostate cancer becomes more prevalent.

Men of African descent have higher prevalence and increased risk of developing prostate cancer. Men of Asian descendent have the lowest risk.

The genetic risk, prostate cancer in the family, indicates increased risk for the development of prostate cancer as well.

According to the Canadian Cancer Society, prostate cancer is the most common cancer among Canadian men and the third leading cause of cancer deaths in Canada. Every day, 65 men are diagnosed with prostate cancer. Every day, 11 men die from prostate cancer. In a lifetime, 1 out of every 8 Canadian men will develop prostate cancer, and out of those, 1 in 28 will die. The 5-year survival rate thankfully is 96 percent.

Diets high in animal fat (with milk and meat leading as culprits) and sugar, excessive weight, and anything irritating the urinary tract have been linked to the development of that cancer.

In an earlier chapter we discussed zinc for the dink. Now we add it in for the health of the prostate. Foods such as sesame seeds are rich in zinc, a mineral essential to the health of the prostate. According to a study in the Indian Journal of Urology, men with either BPH or prostate cancer have lower levels of zinc in their bodies, sometimes up to 75 percent lower than healthy prostates.

Zinc that comes from food is easier to absorb than zinc supplements. Help your body by snacking on sesame seeds. Or try oysters, adzuki beans, pumpkin seeds, and almonds, which are all high in zinc.

Salmon is rich in **omega-3 fatty acids**. These are healthy fats that can protect you from cardiovascular disease, cancer, and rheumatoid arthritis. Fatty acids also help in the synthesis of prostaglandin. Fatty acids deficiency may lead to prostate problems, according to a study published in the Alternative Medicine Review. If you're not a fan of fish, you can get your omega-3s from walnuts, ground flax seeds, canola oil, and kidney beans.

Interesting enough, a longitudinal study done in Sweden found that men who never ate fish were 2 to 3 times more likely to develop prostate cancer than those who regularly consumed it. It seems that it is the omega-3 in fish that plays a role in the health of the prostate.

A study published in The Journal of Urology showed that Asian men have a lower risk of developing BPH than Western men. One possible reason is that Asian men eat more soy. Soybean isoflavones have been linked to a lower risk for an enlarged prostate, according to a study published in The Prostate. Eating more soy might even reduce the risk of developing prostate cancer. For other sources of soybean isoflavones, try low-fat soymilk, tempeh, roasted soybeans, soy yogurt, and meat substitutes made with soy.

But be careful with soy, guys, because it has been found to create a change in the estrogen to testosterone levels, causing an imbalance that can result in a decreased ability to "get it up."

However, in the North American society, the over-processing of soy has significantly reduced the benefits of it, more so than what one would find in places where the soy remains in its more original form.

Vitamin C is an antioxidant that might play a role in fighting BPH. Not all vitamin C is the same, however. According to the Mayo Clinic, only vitamin C obtained from vegetables lowers your risk of an enlarged prostate. Fruits don't offer the same benefit.

Bell peppers contain more vitamin C than any other vegetable. One cup of raw bell peppers contains 195 percent of your daily requirement intake of vitamin C. Other vegetables to try include broccoli, cauliflower, kale, and Brussels sprouts.

Tomatoes are rich in **lycopene**, the bright carotenoid that gives tomatoes its red color. Lycopene may lower the risk of developing prostate cancer. It can also help men with BPH, according to the National Cancer Institute. Lycopene also helps lower the blood level of an antigen, a protein connected to prostate inflammation and BPH.

Tomatoes and tomato products (such as tomato sauce and tomato juice) are the best source of lycopene. You can also get this carotenoid from watermelon, apricots, pink grapefruit, and papaya.

Remember the testicle tree? Avocados are rich in **beta-sitosterol**, a plant sterol. According to the Cochrane Database of Systematic Reviews, they found that beta-sitosterol can help reduce symptoms associated with BPH. Men taking beta-sitosterol supplements ended up with better urinary flow and less volume remaining after peeing. Beta-sitosterol can help strengthen the immune system. It can reduce inflammation and pain, as well. Besides avocadoes, other foods rich in beta-sitosterol include pumpkin seeds, wheat germ, soybeans, and pecans.

Increasing evidence suggests that a diet high in meat—particularly if it's well-done—may be associated with an increased risk of developing prostate cancer. This may be due to heterocyclic amines (HCAs), carcinogens found in cooked meat that have been implicated in the development of several cancers.

HCAs are compounds formed during high-temperature cooking, such as broiling or grilling. Some studies suggest that red meat and processed meat such as beef, pork, lunch meats, hot dogs, and sausage, may be associated with an increased risk of developing prostate cancer.

The World Cancer Research Fund 2008 report suggests that eating large amounts of dairy products may increase cell proliferation in the prostate, which can lead to prostate cancer. According to studies published in The Journal of Nutrition, drinking whole milk increases the risk of progression to fatal prostate cancer, while skim and low-fat milks increase the risk of low-grade stages of the disease. If you don't have to drink milk, stay away from it. It you do drink milk, find the organic milk and limit consumption to low fat in order to give your prostate the best advantage it can have.

Although red wine may do wonders for your heart, early research suggests that alcohol may be harmful to your prostate. Using data from more than 10,000 men participating in the Prostate Cancer Prevention Trial investigators found that heavy alcohol drinkers (>50 g/day) and regular heavy drinkers (≥4 drinks/day on ≥5 days/wk) were twice as likely to be diagnosed with advanced prostate cancer as compared with moderate drinkers. Most of the heavy drinkers in the study drank beer, so researchers were unclear as to whether wine and liquor carry similar risks.

Saturated fats are found in meat, dairy, salad dressings, baked goods, and processed foods. Although saturated fats have been linked to heart disease, the association with prostate cancer is being determined. Researchers at MD Anderson Cancer Center examined the association between saturated fat and PSA levels among patients who had undergone surgery to remove prostate tumors.

PSA, or prostate-specific antigen, is a protein produced by the prostate. Elevated PSA levels in the blood can indicate prostate cancer in some men. However, numerous studies and comparisons have discovered that the PSA may not be a good indicator, as it has been known to give both false positive and false negative results.

A false positive means your tests indicate you have the disease when you really don't. A false negative means the test indicates that you don't have the disease when you really do.

There is also a link to the role of exercise to prostate health:

Italian researchers randomly assigned 231 sedentary men with chronic prostatitis to one of two exercise programs for 18 weeks: aerobic exercise, which included brisk walking, or non aerobic exercise, which included leg lifts, sit-ups, and stretching. Each group exercised three times a week. At the end of the trial, men in both groups felt better, but those in the aerobic exercise group experienced significantly greater improvements in prostatitis pain, anxiety and depression, and quality of life

Instead of focusing on specific foods, holistic nutritionists, MDs, and researchers recommend an overall pattern of healthy eating. Such as:

- Eat at least five servings of fruits and vegetables every day. Go for those with deep, bright colour.
- If you can, avoid bread all together. However, if you eat bread, choose whole-grain bread instead of white bread, and choose whole-grain pasta and cereals.
- Limit your consumption of red meat, including beef, pork, lamb, and goat, and processed meats such as bologna and hot dogs. Fish, skinless poultry, beans, and eggs are healthier sources of protein.
- Choose healthful fats, such as olive oil, nuts (almonds, walnuts, pecans), and avocados. Limit saturated fats from dairy and other animal products.
- Avoid partially-hydrogenated fats (trans fats), which are in many fast foods and packaged foods.
- Avoid sugar-sweetened drinks, such as sodas and many fruit juices. Eat sweets as an occasional treat.
- Cut down on table salt, selecting Kosher or Celtic salt instead.
- Limit the use of canned, processed, and frozen foods.

Watch portion sizes. Eat slowly, and stop eating when you are full. See **Appendix C** for a great Salmon Salad recipe to help keep the prostate healthy.

Penis Protocol 27:

To ensure that you are giving your prostate all the health it can get, eat plenty of vegetables, especially those with vibrant, deep colour. Focus on eating fish, chicken, beans, or eggs as a source of protein. Healthy fats, such as olive oil, nuts, avocado are a great idea. Avoid sweetened drinks, including fruit juices. And if you are using salt, use Kosher or Celtic Sea salt.

Section VIII – Penis Myths

28) 16 Myths Busted About the Penis

Myth: Shoe size dictates penis size.
Research refutes this.

Myth: The penis has a mind of its own.
Well, no. But letting the little head control the big head usually leads to problems.

Myth: No way you can have an orgasm without ejaculation.
That is false. Look at the Taoist philosophy, and the Tantric practice as well.

Myth: The penis is a muscle.
This is false. The penis is made up of two sacs that fill with blood. This is what is responsible for the erection.

Myth: Size matters.
Depends on who you ask; could be a blessing or a curse.

Myth: Masturbation is not a common thing.
False: A study out of the University of Chicago found that 61 percent of men masturbate. And men who masturbate continue to do so throughout their life.

Myth: The size of a flaccid penis is related to the size of an erect penis.
Well, yes and no. The penis can only grow as long as it can be stretched. Divided into two categories, there can be a "meat penis" where the flaccid penis is less than 2 times the length, and the "blood penis," in which the length is greater than double the flaccid length.

Myth: Penises can't be broken.
False: Oh yes they can. Sex-related trauma is the usual culprit.

Myth: No one is allergic to Semen.
Yes, it is possible that someone could be allergic. But it would be an allergy to the proteins in the semen and not the semen itself. Some numbers indicate that this sensitivity affects some 20,000 to 40,000 Americans. Some men are even allergic to their own cum.

Myth: Technology has no influence on Erectile Dysfunction.
False: Look at penile implants, transplants, and vacuum traction.

Myth: A lot of men think their penises are smaller than average.
True, they think that, but it's as a result of viewing porn, or those with a big belly. But what is false is that the majority who think their penises are smaller are actually are within the normal range.

Myth: Americans have bigger penises than Canadians.
False: American 12.9 cm (5.1 ")
          Canadians 13.9 cm (5.5")

Myth: Men want to sleep with their friends' wives.
FALSE: If you're worried about adultery within your friendship circle, this new research may ease your concerns. A recent study (2012) from the University of Missouri found that male testosterone levels drop when interacting with the spouse of a close friend. Whereas, if they are interacting with a potential sex partner, their T levels generally will increase.

MYTH: Masturbation can make you go blind.
Masturbation cannot make you go blind, deaf, nor will it stunt your growth.

MYTH: Masturbation makes you grow hair on your palms.
Have you ever seen someone, anyone, with hairy palms, given that 61 percent of the male population masturbates?

MYTH: Masturbating too much makes you sterile.
False: Sperm is being produced every day. Your physiology and what you consume daily would contribute more to fertility or sterility.

Penis Protocol 28:

Myths are myths. Find out the facts and consider the sources.

Section IX-Extra Penis Stuff & More

29) Masturbation

Choking the chicken, spanking the monkey, or beating off, as the guys in high school used to call it. Regardless of what you call it, it is still masturbation and most men do it. It is pleasurable, healthy, and an integral part of a man's sex life. It is not without its perils, though.

With masturbation beginning at very early on in childhood, some people speculate that men masturbate more than 4,000 times in a lifetime.

Things that are great about masturbation:

1. Keeps your blood flowing to the genitals.
2. More oxygen to the brain with orgasm.
3. Natural painkiller.
4. Natural sedative for some.
5. Gives the mind a break from stress (stress reliever).
6. Good practice for premature ejaculators to know when to stop.
7. Can't get pregnant from masturbating alone.

Things not so great about masturbation:

1. Can cause sexual exhaustion.
2. May lead to youth impotence in teenagers.
3. May become addicting and become a choice over being with a partner, thus less intimacy.
4. A mental health risk if one becomes addicted where he prefers to stay in and masturbate vs. being out in the real world socializing.

5. Over masturbation may lead to decreased T and, with that, decreased sex interest.
6. Man may get hooked on a certain amount of pressure or technique when masturbating that is difficult to reproduce.
7. If masturbating to pornography, that person may not be able to participate with a partner, and thus may need the porn to "get off."

To know yourself intimately, to know how you like to be stroked and touched, is very important for satisfying sex. If you are connected with yourself, have confidence in how you know you like to be touched, then you can respectfully share that with your partner. That partner may be very good at touching you but through respectful communication you will get touched exactly as you want.

In the olden times, numerous devices were concocted to stop boys and men from practicing this "sinful behavior" of masturbation. For example, a penis ring with razor sharp teeth, with extreme pain as a motivator, was created to deter any penis expansion.

If you masturbate, masturbate safely. Do not stick your penis in anything that can break, or something that can break your penis or get it stuck. If you are masturbating, see if you can, over time, gently lengthen your penis through stretching during masturbation.

Some couples use masturbation as part of foreplay. Some find masturbating with a partner to be very erotic. However, if masturbation is taking priority over sex with your partner, that is a situation that needs to be addressed. If this is happening, maybe there are some emotional issues blocking your relationship that need to be addressed before the relationship can go to the next level.

If masturbation is killing your relationship intimacy, seek professional help before it gets too far out of control.

Penis Protocol 29:
Masturbation is a normal part of life for a man. Everyone has done it; it is not taboo.

30) Addicted to Porn?

Porn rewires the brain. There are brain centers that love to be excited to release the feel good hormone. Porn stimulates this feel-good hormone.

To address the issue of porn, we first have to talk about the brain and its neuroplasticity and how it might get rewired.

This neuroplasticity is the ability of the brain to compensate, and then to change neural structures that are the result of repeated neural stimuli (i.e., a habit). The habit then feeds the reward centers in the brain that are stimulated when the brain chemical dopamine is released. Dopamine is the pleasure/happiness chemical for the brain.

Here is how the addiction may start:
Dopamine is released with pleasurable stimuli – people, places, or things. Once dopamine is released, that person may start seeking out those people, for example, who agree with them, because that type of interaction releases dopamine. The individual may start searching out things that are novel or unique, which further releases dopamine. If it is a sex-based search, then the person seeks sexual arousal, which is the big hit of dopamine once the search pays off. When the effects of the dopamine wear off, more dopamine is needed and the search is on again. That is the addiction.

Similar to what a drug addict experiences, the "hit" of porn creates a dopamine release, and once experienced, the desire to regain that feeling surfaces. So, more porn is consumed. That feel-good chemical then becomes increasingly sought after. Thus, the addiction.

This repeated neural stimuli to the brain, the stimulated reward centre, the feel-good centre, is what some feel lead to the porn addiction. It might start innocently enough, if those words could apply in this context, with one or two viewings. But, just like a bad car accident, you just can't help but look, and look again. Although you probably may not like what you see, you just have to look. You crave it just for a moment, which turns into an hour. You chastise yourself for doing it, but the next day you can't help yourself. You do it again. This is what porn can do.

When the neural stimuli become ingrained, men might start thinking about porn the moment that they start thinking about turning on their computer. This would be similar to the drug addict's thrill of walking past the corner where cocaine was tried for the first time.

The following is not a new thought, but porn is shifting the way relationships are initiated and evolved. As men watch porn, they make self-judgments on their penis size and technique ability. If teens are watching, porn may create misguided thoughts as to how people want to be treated in sexual situations. Porn may make things too "easy" for sex, as you get to have many virtual sexual partners simply by watching porn without the complications or complexities of a relationship.

Occasionally, porn is the wanted lover, and the flesh and soul partner no longer becomes attractive. The new lover is not demanding, is always "fresh," novel, and alluring. At times, men are not able to get it up or get it off – without first watching porn – before, during, or after sexual experiences with a partner.

Just as a drug addiction can create havoc and ruin a career, family, finances, and friendships, so too can porn. The feel-good nature of this dopamine addiction is so strong it can cover up where and when we might be vulnerable, such as in depression or loneliness. This then leads to the trap of needing

115

more porn for the dopamine fix, which leads to more porn viewing.

So what happens when you try to withdraw from porn and try to break this bad habit? Porn has created elevated need to supply the dopamine centres of the brain. As such, it has rewired the brain, creating the neural pattern for porn.

The chemical shift and then the new neural firing pattern now go into "detox" and withdrawal when you abstain from watching porn. Not surprisingly, physical symptoms crop up, including insomnia, depression, anxiety, restlessness, emotional instability, headaches, concentration problems, brain fog, and loss of libido.

"What can be done so that I stop watching porn?"

- Admit to the addiction and understand without judgment that it is neither good nor bad that you watch, but that it is teaching you something about your life that you are avoiding. Maybe the relationship you are in. Maybe your sexuality choices.

To help break the addiction, try this exercise. Write down in list form the 50 benefits or drawbacks to the following:
- What are the benefits of watching porn?
- What are the drawbacks of watching it?
- What are the benefits of not watching porn?
- What are the drawbacks of not watching it?

Fifty benefits seems like a lot, however, in order to exact a change in brain and body chemistry there has to be a conscious shift. If there were fewer benefits to write out, this exercise would be an intellectual one and guys will use that as an out not to do the work. You have to do the work to get the behaviour change. That many, the 50 benefits, or more, will exact that shift. This will help you see that there is balance in

116

everything you have chosen to do, and it will help you start to break the addiction.

To keep you on track and minimize the time spent on porn, try the following:

- Every night before turning in, write a list of all those high priority things that you want to accomplish tomorrow. Keep yourself on target.
- Have an accountability partner to whom you send your high priority list to and your daily accomplishments of that list.
- Give yourself a reward for accomplishing your list.

Penis Protocol 30:
Porn can be addictive. Find your way to break free. Focus on what you want to accomplish to grow your life, and keep yourself on target.

31) Piercings

Body piercing is one of the oldest, and some claim, most interesting form of body modification that exists.

Going back through history, evidence shows that Egyptian Pharaohs pierced their navels as a rite of passage; Roman soldiers pierced their nipples to show manhood; Victorian royalty (both male and female) pierced their genitals and nipples. When royalty did this, it was a way for them to be seen as courageous and virile. History is rife with these claims as people have adorned, decorated, and altered their bodies in many different ways.

You can pierce anything in your body, including your penis.
Yeah, we have seen ear, nose, lip, belly button, nipple, eye brow, and tongue piercings. But piercing of the penis? Penis body piercings are done for aesthetic reasons and also to increase the intensity of a sexual experience.

There are books written about piercing and adorning the penis with metal; history has it as far back as the Kama Sutra of 400 A.D. and also to the Ancient Greeks. Some of these current trends in penis piercing also have their roots in ancient tribal customs.

To review the anatomy, the penis has five sections: the base; the shaft; the foreskin that in some intact males covers the head of the penis; the glans (the head); and the meatus (the opening for semen and urine).

Just to summarize the different styles of piercing briefly:

- Ampallang: horizontal piercing through the glans of the penis.
- Apadrayva: vertical piercing through the glans.
- The Prince Albert: through the underside of the glans.

The Prince Albert piercing is named after Prince Albert, who was the husband of Queen Victoria of England. He was reputed to have had this piercing done prior to his marriage to the queen around 1825.

- The reverse Prince Albert: through the topside of the glans.
- Dydoe: Through the rim of the glans.
- There is also foreskin, frenum, Lorum, and Hafada piercing as well. There most likely are many more, but you get the idea.

Deciding on a genital piercing is a profoundly personal choice. When having these procedures, one must also be aware of the time of healing for any of these adornments:

- For the Prince Albert healing, times are from 4 to 8 weeks.
- The Apadrayva heals in 4 to 6 months or longer.
- The Ampallang heals in 4 to 6 months.

It is really tough to say what will happen during those time periods of healing. For instance, what will happen when the morning erections surface – will it create pain or discomfort?

How do you keep things clean and infection-free, and how do you move around post-surgery to avoid banging the metal and swollen penis on something?

Do you want to go 4 to 6 months without sex?

In the culture of today, these ancient practices have been revived, making body piercing a form of self-expression, and an attention-grabbing and exciting way of transforming the body.

Penis Protocol 31:
Before you decide to pierce any of your nether regions, investigate all aspects, understand infection rates, pain involved, recovery times, and ease of self-care.

32) Tattooing Your Willie

Getting a tattoo is commonplace these days. However, getting a tattoo on your willie is another story altogether. People who get tattoos on their wiener are into body modifications, and sometimes they make what some would call radical decisions to be different.

Knowing that your sausage has a lot of nerve endings to it, particularly the glans/head of the penis, there is reportedly quite a lot of pain associated with tattooing the tip of your wanker. The Johnson is not erect while doing this and the tattoo artist holds the knob firmly and pulls it taut for proper needle insertion.

Things that could go wrong with tattooing your rod:

- Infection.
- You could develop a permanent erection.
- Nerve damage.
- Skin damage.

Once you have finished getting the tattoo on your willie, you have to watch closely for any signs of infection. You must keep the area clean and dry. After healing, use creams with proper nutrients to aid in future healing and to ensure skin health.

Penis Protocol 32:
Understand the pros and cons of deciding to tattoo your penis. From cost, to ease of application, to pain levels, to rates of infection, it's best to be well-informed before drawing permanently on your schlong.

33) Toys for Boys

Since this is a book called Penis Protocol, this section is not about Ski-doos, Sea-doos, motorcycles, four wheelers, or other type of big toys. This section has to do with sex toys for men.
Some guys may immediately jump to the next section, thinking toys are only for women and gay men, but they are not. Toys are just another way of finding out about how your body works and other ways to bring sensations and sexuality to a higher level of awareness. Using toys won't transform hetero men to homo men or vice versa.

One gadget, which the medical profession developed for use, was a massager for patients suffering from certain prostate conditions. From this prostate massage, a reduction in the fluid in and around the prostate gland occurred. But because of the intense feeling within the man that happened, it became so popular that the spin-off was a very popular sex toy.

There is an unbelievably large array of toys for men. Including but not limited to:

Butt plugs of all shapes and sizes, anal beads, male g-spot vibrators/stimulators, pocket pussy, glow-in-the-dark anything, cock rings, electric gloves, ball vibrators, penis vibrators, remote control vibrators with a large range in activation distance, masturbation gloves, turbo suck, and an extensive array of designs in handheld (no pun intended) masturbation units.

Ever wondered what some of these do?

The butt plug has its benefits to the penis. During orgasm, the pelvic floor muscles, and the prostate contribute to the ejaculation. During sex, when the orgasm occurs, the anal muscles contract around the butt plug, making the contraction stronger and feel harder with more intensity and duration of the orgasm.

If that is true and really happens, it be interesting for you to find out the accuracy of that statement for your own sexual experiences.

Anal beads on a string are sex toys for men and women. Various strategies are employed for your enjoyment. After insertion into the rectum, the beads are pulled out slowly one by one, either during sex to heighten orgasm, or solely for anal stimulation. But it becomes a problem if the string breaks.

Cock rings are placed around the base of the penis. This keeps the blood from draining out of the penis, thus prolonging penetration and ensuring rigidity. There are vibrating cock rings, as well. The vibrations will be felt up the shaft of the cock, and the recipient will also receive the vibrations deep within them. If female, the vibration at the clitoris will be notable. Cock rings are to be removed if your member starts to turn red or blue. Usually leave it on only for 20 to 30 minutes.

The pocket pussy is just that, a hand held instrument that feels like the real thing. It's a tube made of various materials, battery powered or hand powered, designed to either do all the work for you and/or help you feel as if you are in a real pussy.

Know that these things are out there. I am sure you would be surprised how many guys in straight or gay relationships are using them.

Cleanliness goes without saying, but there are a couple of other things that need to be addressed with sex toys...that is the possible addiction to them, and how easily that can happen.

Similar to the dopamine rush of porn, electric toys may supplant the real thing and create a craving for the use of the toy. Eventually, ejaculation or orgasm may only occur with the toy and not be easily attainable, if at all, with a real flesh-and-blood partner.

Another thing that came about during the research was the concept of desensitization. The overuse of handheld masturbation devices can create diminished sensation in the head of the penis, and possible nerve damage with overuse as well. The nerves to the glans will eventually return to their previous sensitivity, but that time period in which your penis is desensitized can be unsettling emotionally, though it can most likely be prevented with moderation.

Then there are the sex dolls. The blowup kind off the shelf, or the ones you can specially design and order. With the special order, you define and decide on all the specs – breast size, vaginal opening size, anal entry ability, hair, and eye and nail colour. These dolls have life-like skin and life-like lips. They dolls have adjustable vaginal openings so that if you have erection difficulties, you can change the opening size of the doll and actually put your flaccid penis in and bang away on her, hard or soft. You name it, you can order it.

Penis Protocol 33:

Without judgment there is only a choice of using these sex toys or not. Either choice is okay. No one is wrong and no one is right for using them, thinking about them, buying them, or trying them once and deciding they are or aren't for them. It is your body to love, to get in touch with, and understand.

Section X - Getting Up in Age

34) Aging, Sex, and the Penis

When you were younger, you could get a hard-on with the breeze blowing. This instinctive, automatic part of human behavior is one that has complex biology associated with it. This biology and the associated physiology changes over time.

As was presented earlier in the Penis Protocol, there are a complexity of systems that play a role in male sexual arousal, from brain firing, hormone releasing, nerve synapses doing their thing, and blood vessels opening and closing. But over time there is a gradual, almost unrecognizable, slowing of the processes in the male body. This is when the testosterone is dropping. With that T drop, a reduction in sex drive occurs, causing a lowering in erection strength, orgasm intensity, and a lowering in sperm volume.

There are 6 separate divisions to the sex act: sex drive, arousal, plateau, climax, return to flaccid state, and the refractory or reloading period. Yes, they can be seen as individual segments. However, they are really a continuum, without separation. Regardless, almost all divisions are affected by aging.

1. Libido/Sex Drive

Starts in puberty, as testosterone levels begin to rise. It is into adulthood that the levels of testosterone fluctuate somewhat around a mean rate. It is not until age 40 or so that the levels start to drop at a rate of 1 percent a year. As that starts to happen, the libido starts to drop as well. However, even with the lower levels of testosterone, most men make enough testosterone to maintain a sex drive throughout life.

Sex drives start as a result of testosterone's effect on the brain. With age, the nerve cells of the brain become less hormone-responsive. The thought to do it is still there, but there may be less responsiveness with age.

2.   Arousal

Whether it is gourmet sex or junk sex, each must begin with arousal. That arousal could be any one of the 6 senses, in combination with visuals or thoughts.

Part of the brain called the hypothalamus takes information from the senses and thoughts and coordinates it with the neural connections travelling to the pelvis via the spine. From the skin of the penis and other erogenous zones, there is an information loop directly to the pelvic nerves that bypasses the brain. This is all done without conscious awareness.

This arousal is when the blood rushes into the penis, using the neural and chemical signals described earlier, creating the erection. In the area of arousal, many things start to diminish, including: penile blood flow, penile responsiveness, stiffness, and nighttime erections.

As aging occurs, there is apparent shrinkage in length and in girth of the penis. This has to do with possible plaque build-up within the vessels of the penis, impairing blood flow. The other is the build-up of scar tissue, not allowing the tissue of the penis to be as elastic as it once was. Less elasticity means smaller erections.

Also, when the penis shrinks, so does the size of the testicles.

3.   Plateau

This is the phase leading up to ejaculation that lasts 30 seconds to 2 minutes. Blood flow increases around the body, particularly the face, testicles, and penis. Along with this, blood pressure increases. This is the phase when the prostate and seminal vesicles secrete the pre-ejaculatory fluid. Aging has no effect on the plateau phase.

4.   Ejaculation
For most men, ejaculation is the pleasurable sensation of orgasm. This is the phase when semen is propelled or pushed out of the penis through forceful contraction of the testicles and prostate. Simultaneously, the neck of the bladder contracts so that the semen goes out instead of into the bladder.

With age, the musculature pushing sperm out becomes less intense. Thus, ejaculation is slower and less urgent, with less sperm count and less sperm volume. The colour of the sperm also seems to change with age.

5.   Detumescence
This is the time when the penis returns to its flaccid state. The opposite of erection occurs during detumescence when the arteries narrow, offering less blood. The veins also dilate, draining the blood from the penis. This return to flaccid state can occur post-ejaculation or can occur when there is a disruption to the mental state.

6.   Refractory Period
Lasting 30 minutes in young men to 3 hours in older men, the penis is not able to respond to any stimuli during this phase.

Penis Protocol 34:

Sexual Fitness at any age is possible with the proper food, exercise, mental outlook, flexibility, and strength. Regardless, sex is a fundamental component of a healthy lifestyle, and most men remain interested in sex as they age.

35) Boosting Your Testosterone

There are healthy, natural ways to boost your T levels, and a quick medical way to increase your T levels. This chapter will focus on the natural ways to boost up the male sex hormone.

In other sections, zinc was indicated for the dink and for the prostate. To boost up levels of Testosterone, zinc is definitely needed. Natural ways to boost zinc through food choices would be to eat mushrooms, beans, chicken, chocolate, various types of nuts, pumpkin and squash seeds, spinach, beef, lamb, and oysters.

Foods that lower estrogen are important to consume. Foods such as broccoli, cauliflower, cabbage, kale, and bok choy are some examples of cruciferous vegetables, important for your maintenance and/or growth of T levels.

Gain more lean muscle mass through whole body exercises. See **Appendix J** for those exercises and a sample workout program. If you are trying to optimize testosterone, stay away from or minimize the level of booze you consume. As stated elsewhere in the Penis Protocol, excessive consumption of alcohol can have a feminizing effect on the male body. Stress can be detrimental to your health. Stress can cause production of cortisol that can diminish your lean muscle mass, and thus affect your testosterone.

Avoid the junk foods and processed carbs, and settle for regular meals and snacks of low-fat proteins. Once thought of as taboo to eat, the yolk of an egg has the cholesterol necessary for the production of testosterone. So the whole egg is worth eating. As always, fish – a low-fat, high protein meal – is a great source for testosterone building blocks and is low in saturated fats. See **Appendix J** for a sample menu to boost your testosterone.

Penis Protocol 35:

Boost your testosterone with whole body, big-muscle-group exercises and with wise choices of food.

36) Male Menopause

When you think of hot flashes, you probably think of a woman in menopause. But believe it or not, men can also experience the discomfort of flushing and sweating from hot flashes. In fact, according to Harvard Medical School (HMS 2009), although fewer men experience this condition than women do, some men can find hot flashes just as troubling as middle-aged women do.

Menopause in women is a result of a dramatic drop in hormones among aging women. For men, male menopause is a result of a progressive decrease in testosterone levels with age. Then with the low testosterone, changes start to happen:

- **Changes in sexual function.** This might include erectile dysfunction, reduced sexual desire, fewer spontaneous erections — such as during sleep — and infertility. Your testes might become smaller as well.

- **Changes in sleep patterns.** Sometimes low testosterone causes sleep disturbances, such as insomnia or increased sleepiness.

- **Physical changes.** Various physical changes are possible, including increased body fat; reduced muscle bulk and strength; and decreased bone density. Swollen or tender breasts (gynecomastia) and loss of body hair are possible. You may also experience (though rarely) hot flashes, less energy, difficulty in engaging in vigorous physical activity, the inability to walk 1km, or the inability to bend or stoop.

  - **Psychological/Emotional changes.** Low testosterone might contribute to a decrease in motivation or self-confidence. You might feel sad or depressed, have trouble concentrating or remembering things, or experience low energy and fatigue.

With the possible onset of male menopause, it is very important to continue to boost your testosterone with proper food, exercise and rest.

Reducing night sweats can be done naturally through a number of therapeutic approaches, including acupuncture, homeopathy, and naturopathy. One thing that has reduced the number and volume of night sweats in both men and women is the elimination of spicy foods, chocolate, wine, beer or spirits, and ice cream from the evening meal and post-dinner snacks.

Penis Protocol 36:
Male menopause is a real event for men. Just know that it will pass and, through dietary and lifestyle adjustments, you can soften the symptoms or eliminate them altogether.

Section XI – Be Self-Aware

37) Things That Can Go Wrong with Your Package

This section makes sure that you are aware of things that can go wrong with your unit. Don't be paranoid, but be aware.

Below are some examples of disorders that affect men down there (disorders of the bag, balls, or ball material). Conditions affecting the bag contents may involve the balls, ball material, or the bag itself.

- **Testicular trauma.** Even a mild injury to the testicles can cause severe pain, bruising, or swelling. Most testicular injuries occur when the testicles are struck, hit, kicked, or crushed, usually during sports or other trauma.
- **Testicular torsion** - is when one of the testicles twists around, cutting off the blood supply. It is also a problem that some teen males experience - although it's not common. Surgery may be needed to untwist the cord and save the testicle.
- **Undescended Testicles** – Also called Cryptochidism, occurs in 30% of premature babies and 4% of those born at term. In half of those with testicle undescended, most resolve within 6 months. If not descended by then, intervention is necessary to avoid fertility issues and potential health problems later on in life.
- **Varicocele.** This is a varicose vein (an abnormally swollen vein) in the network of veins that run from the testicles. Varicoceles commonly develops when a boy is going through puberty. A varicocele is usually not harmful, although in some people it may damage the testicle or decrease sperm production.

- **Testicular cancer.** This is one of the most common cancers in men younger than 40. The cure rate is high when detected early. But if not found early enough, testicular cancer can spread to other parts of the body.
- **Epididymitis** is inflammation of the epididymis, the coiled tubes that connect the testes with the vas deferens. It is usually caused by an infection, such as the STI chlamydia, and results in pain and swelling next to one of the testicles.
- **Hydrocele.** Usually painless, a hydrocele occurs when fluid collects in the membranes surrounding the testes. Hydroceles may cause swelling of the testicle. In some cases, surgery may be needed to correct the condition.
- **Inguinal hernia.** It looks like a bulge or swelling in the groin area. It occurs when a portion of the intestines pushes through an abnormal opening or weakening of the abdominal wall and into the groin or scrotum.

Disorders of Penis

Disorders of the penis, from birth or acquired, include the following:

- **Inflammation of the penis.** Symptoms of penile inflammation include redness, itching, swelling, and pain. Balanitis occurs when the glans (the head of the penis) becomes inflamed. Posthitis is foreskin inflammation, which is usually due to a yeast or bacterial infection.
- **Hypospadias.** This is a disorder in which the urethra opens on the underside of the penis, not at the tip.
- **Phimosis.** This is a tightness of the foreskin of the penis and is common in newborns and young children. It usually resolves itself without treatment. If it interferes with urination, circumcision may be recommended.
- **Paraphimosis.** This may develop when a boy's uncircumcised penis is retracted but doesn't return to the

136

unretracted position. As a result, blood flow to the penis may be impaired, and your child may experience pain and swelling. A doctor may try to use lubricant to make a small incision so the foreskin can be pulled forward. If that doesn't work, circumcision may be recommended.

- **Ambiguous genitalia.** This occurs when a child is born with genitals that aren't clearly male or female. In most boys born with this disorder, the penis may be very small or nonexistent, but testicular tissue is present. In a small number of cases, the child may have both testicular and ovarian tissue.

- **Micro penis.** This is a disorder in which the penis, although normally formed, is well below the average size, as determined by standard measurements.

- **Sexually transmitted diseases.** Sexually transmitted infections (STIs) that can affect boys include human immunodeficiency virus/acquired immunodeficiency syndrome (HIV/AIDS), human papillomavirus (HPV, or genital warts), syphilis, chlamydia, gonorrhea, genital herpes, and hepatitis B. They are spread from one person to another mainly through sexual intercourse.

- **Erectile dysfunction** This was covered in an earlier chapter.

- **Urinary Tract Infections (UTIs)** UTIs are more common among women than men. But men still get them, with symptoms that include painful and/or frequent urination.

Urinary tract infections (UTIs) are caused by bacteria that have spread to your urinary system, including:

- The bladder – the organ that collects and stores urine.
- Ureters – the tubes that lead from the kidneys to the bladder.

- Urethra – the tube that carries urine from the bladder out of the body.
- Urinary tract infections should be taken seriously. Severe cases may cause kidney or prostate infection if left untreated.

Common symptoms of urinary tract infections include: burning and pain during peeing; the urge to urinate when the bladder is nearly empty; feeling like you need to urinate all the time, especially at night; difficulty controlling when you have to pee; lower abdominal or lower back pain; blood or pus in the urine; and fever.

Yeast infections in men, are less common than in women, but they can still happen. Men can develop oral yeast in the form of thrust. Men can develop penile yeast, which can turn the head of the penis red, and/or you may develop little blister spots around the head as well. Penile yeast infections are usually a result of sexual transmission in unprotected sex.

Men can get spots on their penises, which for the most part are usually nothing to be concerned about. When the spots are noted however, it would be best to observe and take action. Spots are divided into three categories: plaques, ulcers, and papules.

Plaques are usually not a problem, but some are contagious and easily spread. In the rare instances, penile plaques can lead to penile cancer. Names of these plaques include: psoriasis, eczema, lichen sclerosis, balanitis, and posthitis.

Ulcers can be related to more of the ominous causes, including syphillus, other STIs, and penile cancer. Other significant conditions will also present as ulcers.

Papules are generally not thought to be serious, but should be assessed for deeper causes in men over 50. In the papule category are penile warts, psoriasis, eczema, Fordyce spots, and numerous other conditions leading to papules.

Penis Protocol 37:

If something is growing on your penis; if it is flakey, has ulcers, or some other growth; or if your unit is leaking fluid or smelling funny, take action. Do not wait...take care of yourself.

38) Urine and You

You are standing at the toilet taking a pee. Where are you looking? What catches your attention when you are peeing?

The sound of your urine hitting the water might catch your attention, as may the colour, appearance, frequency, and smell of your urine. Part of your health and well-being is about paying attention to your body – the signs, signals, and symptoms.

Looking pale- that means you are drinking the water you need. Many bathroom visits are the result. Too much water can overload the kidneys and dilute the electrolytes in your body. The amount of water you need will vary day to day based on humidity, activity level, and body physiology.

Too dark- usually associated with dehydration and a sign that your kidneys are producing concentrated urine which has a brownish, iced-tea tinge. If you are dizzy or have concentration troubles, look at your urine. If dark, definitely drink some water to eliminate the possibility of dehydration. If it doesn't lighten up after a few glasses of water, it could be blood causing the discolouration. Bright red is different than the dark colour as this may indicate a problem higher up in the kidney. It could be cancer, an infection, or kidney disease.

Sweet smelling- that could mean a problem with your blood sugar, an overload of which could be excreted in the urine. It may be a sign of diabetes.

Funny smelling- if you have been eating asparagus, your urine will have a funny smell. Garlic too. Different foods can make the urine have distinctive smells. These smells are the result of the food, when broken down in the stomach or digestive system, filtering through the kidney and the smell coming out in the urine. Drinking more water will flush the system and the smell will disappear. If the funny smell remains, you may have an infection or urinary stones; best to get it checked.

Urine normally has a strong smell first thing in the morning, as at this time it is very concentrated.

Bright yellow- is usually the result of taking vitamins. No worries.

A spot of blood- it could only take one drop of blood to change the colour of urine in the bowl. Blood may mean a problem with your prostate, a urinary tract infection, bladder cancer, kidney stones, or perhaps a side effect of taking medications such as aspirin or blood thinners. Also remember that eating beets can colour your urine red.

Always gotta go - if that is the case, you could be drinking too much water, have a UTI, overactive bladder, prostate problems, a neurological disorder, or reaction to medication

A little leakage- might be the result of a prostate problem in which you are not able to control the flow. Weakness in the muscles of the pelvis could also cause urinary incontinence, as can yeast infections, which can lead to the inability to hold the urine in (particularly when coughing).

A burning sensation- you probably have a UTI. A burning sensation is one of the first signs. It occurs more often in women. It could also be a warning sign for prostate trouble.

Something as simple as drinking freshly-squeezed lemon water may be enough to clear out any of these problems.

Penis Protocol 38:

Smells and colours changes could be a normal part of living. Make sure if smells and colours remain that you seek professional guidance.

39) How Anger Affects Your Penis

Anger can affect the function of your penis.

Most people do not make the connection between the onset of their problem and the emotional backdrop to their life. If the emotions are left unresolved, they can become the master of our body and we lose control to them. The danger here is that with emotions, we become reactive and are continually triggered by those emotions. We then are led by what our emotions drive us to do. Men can deny that they have the emotions hidden within them and live on, or as some guys say, " I have dealt with it." That is just talk from the ego. They deny it and bury it in some recess of their mind or body, which will present itself later as some form of disease.

Deepak Chopra, Narayan Singh, Louise Hays and numerous other famous health and wellness authors have written in depth about this topic of emotions affecting our body. Being in the clinic setting for more than two decades, I have seen many clients suffering from both the positive and negative effects of their emotions. I've seen how their bodies show the ravages of their emotions, and unfortunately the emotions start to take over and control their pain, behavior, and mindset.

Research has shown that within 10 seconds of having a thought, either positive or negative, the blood starts to change towards health with a positive thought or away from health with a negative thought. For example, anger is an inflammatory reaction in your body affecting heart rate, respiration, sweat glands, digestion, mental alertness, and more.

Those people who have the highest level of anger have 2 times the risk of developing coronary artery disease and are three times as likely to have a heart attack. Not only that, but

143

prolonged stress of anger can lead to headaches, a depleted immune system, adrenal gland fatigue, stomach and intestinal problems, insomnia, and/or skin problems.

At the clinic with friends or colleagues, someone will mention in passing about their body symptoms, and more often than not, the topic is about their bladder infection. A few questions later usually reveal a long burning anger or frustration in which people are pissed off; they have internalized it, denied it, and let the subconscious continue to hold onto it. Guess where the pissed off went? Right to the bladder that holds the urine.

Chronic anger can alter your hormones, and also narrow and restrict blood flow through the arteries. Knowing that your penis has extensive need for blood flow to hold an erection, narrowed arteries and altered hormone levels are a recipe for lack of ability for a "full salute." The chronic stress of anger can influence the function of the liver and liver hormones. It can also create cholesterol and kidneys issues, both of which are required for adequate levels of T and proper sexual functioning.

There are many strategies to safely and effectively work out your anger. Check Appendix L for anger release ideas.

Penis Protocol 39:

Become a master of your emotions, experience them, acknowledge them, thank them and release, but don't let them run your life.

40) Gender Reassignment

Gender reassignment has been described as a complicated, lengthy, and emotionally and physically draining experience. In this chapter, focusing on what happens purely on the surgical and physical side of things will make this discussion shorter and easier.

Male to female transition: With this surgery, the testicles are removed. The foreskin (if intact) and skin of the penis, plus the arteries and nerves left intact, are used to form the walls of the vagina. The remaining portion of the penis, including the glans, is kept attached after surgical dissection. The glans is used as the new clitoris. If the foreskin is available, a portion of it is used for the inner lips of the vagina, with the scrotal sac used to form the outer lips. The vagina then becomes fully functioning for sex and orgasm.

For female to male transition: This surgery requires more external construction to create a member since the parts are not there, and some other parts must be removed. A hysterectomy takes place, along with removal of the ovaries and fallopian tubes. The clitoris, because of its increased level of nerve sensitivities, is used and enlarged by androgenic hormones to become the head of the penis. Tissues are taken from the thigh or belly to form the penis. A prosthetic inflatable device is used for creation of testicles. The outer lips of the vagina are extended to form the scrotum, into which prosthetic testicles are inserted. Erections take place from a hydraulic pump. The reservoir for the fluid is usually located in the abdomen, and the switch to inflate is located in the scrotum. The same switch, flipped to the opposite direction, is used to drain the penis back into the reservoir, returning the penis to its flaccid state.

Penis Protocol 40:
Just love who you are, and who you aspire to be. It's your life, live it.

Section XII –
Having the Conversations

41) Talking to Your Son About His Penis

What I find in today's society is that men are woefully unaware of their bodies. How can men talk to their children about the beauty of the body when they don't know about or love their own body? That has to change.

If men have no idea how their bodies work, maybe they have a skewed perception passed down from their fathers of what our bodies are supposed to do. Chances are there will be avoidance or a gloss over when trying to educate their children about caring for the male anatomy.

Men have to take care of their own bodies and show their bodies to their children in healthy ways. Hiding your body, or acting ashamed of your body, will create the same perspective in your children. How you deal with and how you heal your body will be emulated by your children. Your children are constantly watching you for clues on how to live, and examples of how to take care of themselves.

Take the time to talk to your son about general anatomy, such as the penis and testicles, and what the body parts do. Keep it light. When they have awareness and know what their body looks like and feels like, they become comfortable in their look and movement; it creates a level of confidence. Boys have to understand that almost all other boys have the same stuff going on with their bodies, and that all dads have gone through it as well.

These are talks that need to be reinforced on a regular basis. It is not like the old days when of one talk, "got it son," and then it's never repeated again. Follow-up conversations to make sure they get it, make sure they are doing what they need to do to understand and take care of their bodies and their anatomy, are a necessity.

Show them and talk to them about peeing standing up or sitting down. Who really cares how you pee? Talk to them about why pee comes out and the importance of drinking water. Stimulate the conversation so that they can talk to you as they grow into their manhood, get pubic hair, have nocturnal emissions, masturbate, and work through relationship issues.

The sex talk that kids get at school is basic. You have the opportunity to discuss lightly or deeply – and with inspiration – the beauty of bodies, the need for respect with your partner, the idea of resonance, plus listening and speaking your mind when it comes to the topics of sex. Engage the dialogue of pornography and how it may be a depiction of how some people have sex, but not how others have sex. Talk about how addictions of any type can be a detriment to life and living. Draw examples and share stories, as these are powerful ways to impress what is important. Speak about respect of self and how loving oneself leads to standing tall in life, being authentic, and facing the future no matter what it brings to you.

Discuss with your uncircumcised children the basic health and cleaning habits under the foreskin. Ask them, "Why is doing this important?"

If you have fear about talking about your penis, a natural part of your body, or if you are shy about showing it in the shower or when changing, what will that create for your son? If you speak about your penis in respectful terms, that it is a special thing to have, that guys have different shapes and sizes, and that it really doesn't matter what other guys have down there, you will create a sense of awareness for them; they will become comfortable and relaxed in their body image. A sense of self-acceptance can result.

"What you have son is perfect exactly the way it is."

Penis Protocol 41:

Open the conversation, develop the bond, and expand your relationship.

42) Talk to your Daughter About Penises

I was in the car one day with my daughter, driving to get new shoes, and she says, "Dad, can I ask you a question?"

"Sure," I replied.

She took a deep breath, & with all seriousness asked:

"Are you going through a midlife crisis?"

Without knowing the background leading up to questions such as this, I answered with a question, " What do you mean?"

"Well," she replied, "I clicked open your computer to do some school work, and there was all this sex stuff...like a lot of sex stuff. Dad, what's going on? Are you going through a mid-life crisis? You have to tell me!"

This is how it came about: she had opened my computer without my permission, and when she did she was hit by probably 25 different pages I had written on sex. I had pages written on orgasm, masturbation, anal sex, sex videos, penis facts, pleasing a woman orally, penis piercings, the multi-orgasmic lover, and sex toys. Most everything you see in this book, she saw it first.

So, carefully, I explained what I was doing, speaking in her value system about what I was writing. However, I learned that day about the unpredictability of the preteen female mind. When I was finished, she turned white. Then in a voice, tone, and attitude that only a preteen could deliver (and those who have heard it before know what tone I mean), she said:

"Dad, OMG, how embarrassing! When my friends find out, they will pick on me and bully me, all because of you... all because my dad wrote a stupid book on stupid penises! Please, please, I beg you Dad, don't do this to me!"

This is when she found out I was writing this book. This is how she and I first started talking about penises in detail.

Why avoid it? Why avoid the conversation? There is a lot at stake here. With so much information available from the Internet that could profoundly influence the future choices of our children, we have to take the lead in this discussion.

I know of a man who has two children. He never, ever let them see him naked or let them into the shower with him when they were little. No one could be in the bathroom with him when he showered or went to the toilet. My dad was exactly the same.

I get the idea of privacy, but children need to see human anatomy to understand it and to discover that there is nothing wrong with the naked body. We have to show our bodies to our children in a healthy and respectful way. As part of those teachings, it is also important to let them know that there are times and places where nudity works and where it might not be appropriate.

Having your daughter understand male anatomy in a healthy way, what it looks like, how it works and why, and what testicles do is a great discussion. Having these talks with your girls allows them to understand the differences in boys and girls, and with your body as an example, what happens when boys grow up. And through this there is no built-in fear of the unknown, nor are there misconceptions of what it is all about. Misconceptions that could come from the Internet or perceptions from friends are important things to acknowledge, affirm, or dispel, if necessary.

Look guys, if you are not comfortable with the topic of penises and erections, that is okay. If you are uncomfortable, find the level that you are comfortable with and talk at that level. Talk about it and gain your confidence in delivering the information. Since your daughter(s) don't know what it is all about, take the opportunity to challenge yourself to work in one more little fact or tidbit about sex or reproduction that you are not comfortable talking. If you deliver the information in a casual way, they can feel comfortable talking about it as well. You need and want to have that level of dialogue so that when they hear or see something that confuses them, they have a trusted source to confide in or from whom to seek knowledge and guidance.

Find out if your child is a visual, auditory, or other style of learner and teach to that strength, in a manner that shows loving authority and respect for them, yourself, and for the topic.

Teach her about guys and girls and how they think differently. Teach about respect for self and what that might look like. Teach her about how you take care of your body, and possible strategies of how she can take care of hers.

Educate about what consent is, what is respectful male behaviour, how to communicate and why, and what to do if something happens that may be scary or of which they are uncertain.

Penis Protocol 42:
Develop and maintain open lines of communication with your daughter about sex, intimacy, respect and consent.

43) Talking to Your Lover About What You Like

There have been plenty of exercises up to this point in the book that have drawn your focus to self awareness, being one with the penis, body awareness, knowing your body, knowing your body cues, and ways to slow yourself down when you get too excited.

So now, you know how you like to be touched, and you want to make sure that you know how your lover enjoys being touched as well. You have learned how to be a master of your sexuality.

When you are having or are about to have sex, you activate all of your senses. When you are in tune with your partner, sometimes her/his body signals or movements give you a great indicator as to the effectiveness of your skills.

For example, when you are providing oral sex to your woman and you are into it, one nonverbal feedback that gives you a clue as to whether she is enjoying it or not is when she places one hand on your arm. That is sign of nonverbal approval, unrelated to her sounds, that she is into what you are doing.

If you are receiving oral sex from your partner and their teeth or suction are irritating or hurting your penis, there is no rule that states you have to take the discomfort until it is over. You can lightly touch their head, and gently indicate that less teeth would be more comfortable. Gentle touch, gentle words, and gentle instructions keep the communications open, energy flowing with intensity, and intimacy intact.

What are you seeing? Are the nipples erect? Is there a flush to the face? Is he getting harder and larger? Is she opening and her lips are becoming more full?

What are you hearing? The breathing? The change in voice tone? The sounds of intimacy...what are they for you? Are their words and body matching up? If not, what needs to be modified?
What are you feeling from your partner? Moistness? Heat? Opening and widening? Vibrations of their pelvis?

What are the smells of your partner and how do they change as you transform from the foreplay of kissing to the arrival at third base? Can you smell the hormones of their arousal?

What is the taste of your partner? Do you notice the change in consistency of their fluids with orgasm?

What is their energy that they give back to you? Are they excited about the idea of engaging in vaginal intercourse, and do they become withdrawn and shut down with the mention of anal sex?

Some people in a consensual moment just want to be taken and taken hard, ravishing or being ravished. Sometimes the only feedback you get is the sound of ecstasy and ravishing, and that is just great. Sometimes your partner is telling you in unabashed language exactly what they want you to do to them, and that is all the feedback you need. Other times it is slow and silent, where all the senses come into play. As you know, there are sexual times when both happen within the same session.

Regardless, touching gently to guide, redirect, reposition, and reinforce what you enjoy is so important for the type of sex you want.

Penis Protocol 43:
Here is the Master secret: Communicate to your partner with loving intent, use all of your senses to engage in sex, gently guiding and with gratitude for your partner being with you at that moment.

44) "It Takes Balls To Check Your Bag!"

That was the tag line of the song written by David Shackleton for the men's movement that I want to get rolling across the country.

A testicular self-exam is an easy way for guys to check their package. Checking your package doesn't mean you see it in the mirror, or scratch around your balls. It takes 1 minute a month to do this check.

Monthly self-testicular exams are important to notice changes in one's testicles. Most testicular cancers are found by men themselves or by their partner; very few are found by a physician. Self-testicular exams allow you to become familiar with your testicles, thus making it easier to notice any changes.

If you do notice any changes, see a doctor immediately.

Although testicular cancer is rare, it is the most common form of cancer in men ages 15-35. The stats are that 1 in 250 men will be diagnosed with testicular cancer.

It's easy to check.

Here's how:

1. If possible, stand in front of a mirror. Check for any swelling on the scrotal skin. It's best to do this when the testicles are warm and descended, as in after a shower or bath.

2. Examine each testicle with both hands. Place the index and middle fingers under the testicle with the thumbs placed on top. Firm but gently roll the testicle between the thumbs and fingers to feel for any irregularities on the surface or texture of the testicle.

3. Find the epididymis, a soft rope-like structure on the back of the testicle. If you are familiar with this structure, you won't mistake it for a suspicious lump.

Thank you to the Testicular Cancer Society.

Penis Protocol 44:

"Grab your balls, Check 'em out. That's what this is all about."

"Check 'em out" music and lyrics by David Shackleton (2014)

Find the entire song lyrics on one of the back pages of this book.

Section XIII – Penis Protocol Conclusion

Taking care of your package, your block and tackle, your rhythm stick, your longfellow, is a must. Taking care of your unit starts with what you think about, what you eat, and what you drink. As a little baby in diapers, you were cleaned, washed, and dried down there so that you would have that area free from rashes and "ouchy" spots. You have the same penis as you did when you were born. It's a little bigger and possibly hairier than before, and with testicles, so now it is your turn to take care of it. You are old enough to clean, dry, and moisturize, and keep yourself out of situations that will create ouchy spots or situations that will give you rashes.

Consider that everything you do in your life has an influence on your unit. Sitting for prolonged times, exercising too much or not enough, posture, food, alcohol overuse causing decreased testosterone, oversexed or undersexed, thoughts, beverages, TSE, environmental pollutants, internal physiology, and so much more.

For many guys, the penis is their manhood. To many guys, their ability to have an erection and participate in sex, whatever that might look like to them, is another one of the components of being a man. What if you took the time for your penis and testicles as if your manhood depended upon it? What if you took the same amount of time taking care of your penis and body as you do sitting on the crapper reading a newspaper? How would your overall health improve if you did that?

You will most likely have your penis for life. How you treat your package is entirely up to you. Your member can be the source of great pleasure, great pain, or neither, so it is your choice, but please choose wisely. This book gives you numerous ideas and strategies to deal with many aspects of your manhood – try some of these out. But remember that in order to exact the change you want, it will take time. So practice and patience is required. Only you know what is right, feels right, and what works for you.

Remember:

Be Original   Be Authentic   Be Extraordinary Every day.

Section XIV - Appendices A to N

Appendix A – Find & Strengthen the PC

Finding the PC muscle:

To locate this pubococcygeal muscle is really quite easy. This first step is to place two fingers behind your testicles, with no pressure. Get the picture in your head that you are urinating and you want to stop the flow. The tissue area where your fingers are placed becomes tense as the urine flow is stopped. That is the PC muscle.

Strengthen your PC muscles:

The purpose of strengthening these muscles is to become completely successful in preventing the sexual energy from being dissipated through ejaculation.

Exercise 1:

This is the first exercise, using a rhythmic contraction relaxation cycle, which you will perform daily. The nice thing here is that you can do this anytime and anywhere.

1. Contract the PC muscle and hold for 3 seconds. Breathe as freely as possible.

2. Relax the PC muscle.

3. Repeat steps 1 and 2 twenty to fifty times.

Exercise 2:

Here is the second exercise that you have to perform daily, immediately after Exercise 1:

1. The prostate, perineum, and anus are the focus of this exercise. **Inhale** and direct your intention to these areas.

2. As you **Exhale**, and at the same time that you contract the PC muscle around the prostate and anus, contract the muscles around the eyes and lips. These facial contractions help contractions of your PC muscle.

3. **Inhale** and relax, releasing all the muscles you contracted on the exhale.

4. Repeat steps 2 and 3, contracting the muscles while inhaling, and relaxing them while exhaling. Do 10 to 50 repetitions.

Start with one set of these exercises per day and build up your stamina. After a couple of weeks of practice, start working these exercises two-three times a day.

There are two typical errors in the practice of these exercises:

1. Contracting muscles other than the PC muscle. The muscles of the abdomen, thighs, buttock, and low back need to stay relaxed. Until you are able to isolate the PC muscle, you initially may contract those muscles as well. Through practice, you can zero in on the PC.

2. Overdoing it on the first few sessions. The PC muscle can get overworked just like any other muscle, so gradually build up to the high number of repetitions. No need to force the contractions; slowly build up over time.

For the muscle to become fit, about three to four weeks are necessary, at an interval in which you have to practice the two exercises daily.

Focusing on the PC muscle area will help you target the muscle during your PC workout. However, if you need some help to find it, remember you can use your finger on that PC area when peeing to make sure you feel the tension when you interrupt the urine flow with a PC contraction.

Fit in this practice of PC strengthening; make it part of your routine. Do it in the shower in the morning, or when getting into bed at night.

After 3-4 weeks of practicing daily the exercises described above, you are ready to add in the following exercise:

Exercise 3:

Same as Exercise 1, but now you increase your time of contraction and have a specified time of relaxation.
1. **Inhale** and focus your attention in the pelvic area.

2. **Exhale,** contract the PC muscle, and keep the contraction for 10 to 15 seconds.

3. Inhale and relax the contraction. Stay relaxed for 5 seconds.

4. Repeat steps 2 and 3, contracting the PC muscle while inhaling and relaxing it while exhaling.

Do 10 to 50 repetitions in one session.

Once again, be patient and build up slowly to the 50 contractions in one session. Remember, no forcing.

Exercise 4:

A good method for checking the firmness of your PC muscle is to hang a towel on your erect penis, then lift and lower it through the contractions of this muscle. In the advanced Taoist practice, the men use weights instead of the towel.

You can do this daily or every second day for 10 repetitions.

Appendix B

How to last longer in bed & finding your P-spot

How to last longer in bed:

- Watch your breathing. Breathe...it slows you down. Breathing makes the energy flow all around and makes your body relax. Holding your breath speeds you up, makes your body tense, and takes you quickly to the end.
- Take a look at what you focus on during sex. If you focus on your partner, you are ignoring your sensations being created, so by staying focused inside your own body you gain self-awareness and have control.
- Pace yourself, take your time, slow down.
- Be in no hurry to bang hard, unless you have practiced this and it is no problem for you. Otherwise, build up to it.
- Understand your cues or triggers that "set you off." Then, while alone or with a partner, practice so that those triggers become less activated.
- Have your partner gently pull on the testicles so that they are moving away from the body. That will help your arousal level go back to pre-critical levels.
- Make sure you are eating well. Not having enough food will lower your energy. You will not have the stamina to get the juice to the motor.
- Focus on being in the moment and keep your thoughts on yourself, your sensations, and how things feel. Worries, unpleasant thoughts, etc., will create a decreased environment for prolonged stamina.
- Find interesting positions that you are less likely to ejaculate from; for instance, the "reverse cowgirl."

- **How to find your male P-spot**

After your partner has trimmed his or her nails, get plenty of lubrication. Place all the lube around the opening and your partner is ready to begin. Make sure the room is comfortably warm as well. A hot bath or shower will also help the tissue relax.

Have your partner start with a light movement with both hands, the movement simulating a spreading motion both on your buttock cheeks and close to the tissue surrounding the anus. This motion starts to warm up and stretch the tissues; have them ready to accept the finger or appliance.

Important: if you are using a toy, make sure that it is either on some type of tether, or has a long handle; when placed into the rectum, the muscles will actually pull the object into the rectum, making retrieval difficult. A trip to the emergency room will make the story of the night for the nurses, doctors, and residents for months to follow. You don't want that.

Your position, as the recipient, is on your back with knees drawn up and slightly outward.

With the lube in place, your partner can now gently trace around the anus for several minutes, helping to relax the receiver before gradually entering with the finger. The finger moving in is done with the pad of the finger, not the tip. Once the finger is in place, it may take a little while for the anal sphincter muscles to relax enough, but once the finger is in, avoid moving in and out. Only move it out to put on more lube.

As the recipient, you have the job of relaxing and letting go of the muscles. Since the normal sensation is to have things moving out of your anus, having something sliding in is rather unusual at first, but just breathe; let the muscles relax and feel the opening.

Once the finger is in, your partner is feeling for a shape and size of a walnut, which is located about 5-8 centimeters from the opening, but on the front wall of the rectum. It is usually found with the palm up while the finger is inserted. Once found, massage the prostate with a "come here" finger movement. The sensations will vary with the intensity of the pressure. The pressure on the prostate can actually inhibit ejaculation, which means these sensations and this type of work can continue for as much time as you choose.

During this massage, the giver or receiver may initiate masturbation up to the point of climax or orgasm.

Appendix C

Recipes - Skin Emollients

Try this little recipe for an emollient that you can prepare at home for use anywhere on your body:
4 tbsp of Beeswax, chopped into small pieces.
¾ cup of olive oil.
¾ cup of vitamin E oil.
8-10 drops of an essential oil of your choice. Lavender oil has anti-fungal, anti-bacterial, and anti-inflammatory properties. It's great for the nether regions. Guys – think jock-itch relief.

In a small saucepan on the lowest heat possible (or you will burn it), melt the beeswax with the oils. Stir. Remove from the heat and let it cool to room temperature. The next step will be to place everything in a variable-speed blender with 1-2 cups of reverse osmosis filtered water. Ensure that the blender has been sterilized.

Avoiding tap water is important due to the nature of impurities, chemicals, and microbes in that water.

There are many devices that you can use to attach to showerheads or taps, or mount under sinks, to purify the water going on and into your body.

While blending, watch the mixture closely to ensure you get the blend to the consistency that you want. Once done, place your homemade emollient into sterilized glass jars. It can be kept for up to 6 months.

The Rocket Pack or The Erection Diet

This is great for during the day at work.

Snack 1: banana, 1 cup of blueberries, handful of almonds.

Lunch: Arugula Salad (one cup) with lemon and olive oil dressing; one breast of chicken cooked the night before; and coconut water.

Snack 2: boiled egg, or ½ cup of hummus; rice crackers; gluten-free bread with almond butter; one stick of celery; one salmon   salad.

Salmon Salad - Great for the Prostate

Yield: 1 single portion
Ingredients:
Arugula - 1/2 lb
Pine nuts - 2 tbsp
Salmon fillet - 4 oz portion
Sun-dried tomatoes - 2 tbsp
Lemon dressing - 1-2 tbsp
Goat cheese - 2 tbsp
Salt and pepper pinch of both

Lemon Dressing

Yield: ~2/3 cup
Ingredients:
Vegetable oil - ¼ cup
Lemon - 1 zested
Lemon juice - ¼ cup
Sugar - 1/5 tbsp
Onion - 1 tbsp
Salt pepper - pinch each
Cider vinegar - 1.5 tbsp

Method:
- Set up both of your pans on a medium-high heat. Add your pine nuts (we are going to toast them). Be careful – they will burn quickly, so keep an eye on them. A golden colour is what you are looking for, nothing darker.
- Take your salmon fillet and give it a little sprinkle of salt and pepper. Lightly oil your pan and set your salmon in it to cook (a barbecue is also a perfect alternative here)
- Cook your salmon for 2-3minutes per side, flipping it halfway through.
- In the meantime, prepare your salad.
- Toss your arugula with lemon dressing, and salt and pepper to your taste.
- Place your arugula nicely in a bowl and sprinkle your toasted pine nuts, sun-dried tomatoes, and goat cheese on top.
- Once your salmon is finished, place it on top of your salad.

Lemon Dressing:
- Gently sauté your onion diced in a pan (you can even

use your salmon pan before you cook the salmon; just add a small spray or splash of oil before the fish goes down).
• In the meantime, combine your remaining ingredients in a blender or with a hand blender – add your onions once they are done.

This dressing keeps in the fridge for ~10 days and is very versatile if you wish to tweak it or use in on a variety of dishes.

Appendix D

Prepping for a manscaping session

- Make sure skin is clean and dry because sweat is the enemy for waxing.

- Have no oil, lotion, or moisturizer on the skin.

- Make sure your body hair is trimmed down. The appropriate length is 1-2 cm so that it can at least lie flat on your skin.

- Find your happy mental spot and relax- a very important step in all of these.

- Avoid any foods or drinks that will jack you up and have you bouncing off the walls.

- Wear loose-fitting clothes to avoid skin rubbing post-wax.

- Know in post-wax that the skin will be sensitive, with possibly some small bumps, redness, and maybe some heat. It will all pass within the first 24 hours.

- Avoid exercise and keep the area clean and dry for 24-48 hours. No swimming, saunas, whirlpools, hot showers, or baths.

- Avoid harsh chemicals, perfumed soaps, scented body creams, sun tanning, or sun beds.

After the work is completed, wear loose-fitting clothes (boxers, for instance) to keep air circulating and the friction on the shaved area(s) at a minimum.

Appendix E   Differences between underwear

Boxers

Boxers have elastic waistbands and wide leg openings like shorts. Some styles feature an open fly, while others have buttons or fasteners that close the fly. The loose design allows air to circulate, but provides no support. Boxers are acceptable as loungewear and comfortable to sleep in. When worn with pants, this style tends to bunch up and ride above the waist. A variety of colors, patterns, and designs appear on boxers.

Briefs

Briefs are a form-fitting style of underwear that has elastic leg openings and a waistband. The fly features overlapping pockets of material for support. The traditional color is white, but some pairs are available in solid colors or depict logos and images. This style provides sufficient support for physical activities. Over time, the fabric may become worn and stretch out. It is best to replace the underwear when it no longer conforms to the body.

Boxer-Briefs

Boxer-briefs combine the coverage of boxers with the support of briefs. The legs extend to the middle of the thigh or higher and hug the body. The overlapping or sewn fly offers support for any type of activity. Boxer-briefs can be worn underneath pants without bunching or displaying lines. Many men prefer this style for its versatility and comfort. Most pairs are available in solid colors ranging from light to dark shades.

Thongs

Thongs are considered a sexy style of men's underwear. They feature an elastic waistband of varying widths, a pouch, and a thin fabric strip for the back. This style will not show lines when worn with pants, but some men consider thongs to be uncomfortable. Thongs offer some support but are not recommended for physical activities because of discomfort and sweating. Solid color options are usually the norm. The g-string is a variation of the thong with an extremely thin strip of fabric in the back.

Bikinis

Men's bikinis are another sexy style of underwear. They are designed like briefs but with a more revealing look. The waist typically sits slightly lower on the hips, allowing for lower cut jeans. String bikinis are a variation with narrow waistbands at the hips. When worn with pants, bikinis may ride up in the back and cause discomfort.

Thermal Underwear

Thermal underwear is form fitting and extends from the waist to the ankles. Some styles come in sets with bottoms and long-sleeved tops. The center is supported with an overlapping fly, and the material is usually insulated. Men typically wear thermal underwear in cold climates where maintaining body heat is essential for comfort and survival. It is common for men to wear this style as loungewear around the home or for sleeping.

Appendix F

10 Steps to stronger erections:

Step 1:

Take the actions you need to decrease the smokes, alcohol, and pop that you consume.

Smokes affect the arteries and blood supply – not only of the penis, but the whole body. Poor blood supply means less blood, and less blood to the penis means no "pokey pokey." Drinking alcohol affects the liver, drops T, and allows the feminizing hormone estrogen to exert more of an effect.

Drinking pop (phosphoric acid within) leads to heart and kidney problems, a decrease in muscle mass, and osteoporosis. If you remember from an earlier chapter, all of these were mentioned in conjunction with low T. With low T, you get low sex drive.

Step 2:

Take a look at the meds you may be taking. They might have an effect on your erection stiffness. The following have been found to reduce stiffness:

Statins – which are designed to lower cholesterol levels by reducing cholesterol levels produced by the liver; an example of this is Lovastatin.

BP meds – such as beta-blockers or diuretics.

Antidepressants – reduce the symptoms of a depressive disorder by correcting imbalances of neurotransmitters in the brain. Examples include Zoloft, Paxil, and Prozac.

Antipsychotic – these work on the neurotransmitters of the brain, allowing communication between nerve cells. An example would be Respiridone.

Benzodiazepines – psychoactive drugs that work on the neurotransmitters used to treat anxiety and convulsions. They are also used as a muscle relaxant and sedative. A classic example of this would be Valium.

H2 Blockers – used for heart burn or stomach acid reflux GERD. Examples of this would be Xantac and Pepcid.

Anticonvulsants – stop or reduce the severity of epileptic or other types of convulsions. Examples are Lyrica and Topamax.

Step 3:

Eat properly. Select foods that are healthy for your arteries. Avoid eating soy as that can increase your estrogen levels and decrease your testosterone.

Avoid sugar in your diet. Sugar will spike your blood sugar, cause insulin to be released. That drops your T.

Foods to boost your hardness would include salmon, fish, blueberries, walnuts, asparagus, and arugula.

Step 4:

Choose sexual positions that keep the blood flowing into the penis during intercourse. That means that during sex, use gravity to keep your erection hard (missionary, doggy style, etc.)

Avoid your partner being in the on-top position to begin with when having sex, as gravity may flow things in the opposite direction and drain the penis of blood.

Try a penis ring, which traps the blood within the penis and does not allow it to flow out as easily.

Step 5:

Be relaxed and have fun. Get comfortable during sex, unless of course you are doing it in a bathroom at a local restaurant...that is a different kind of fun.

Step 6:

Although your penis is not a muscle, you can work out a muscle that will help with your erections. Work out your PC muscle. This will fortify the muscle and will help with stronger erections and stronger orgasms.

See Appendix A for these exercises.

Step 7:

Avoid paper and products that have Bisphenol-A (BPA). It can be found on grocery receipts and in canned products. BPA and other reactions in the body can cause an increase in female hormone estrogen.

Step 8:

Work out with your whole body using compound exercises, which boost your T levels and clear your mental state. Remember that over-exercising can actually lower your T and thus weaken your sexual response. Understand and know what your exercise limit might look like. More exercise isn't always better.

See examples of such exercises in Appendix J.

Step 9:

Get plenty of sleep. Less than 5 hours of sleep a night for several days in a row drops your testosterone levels. With 8 hours or more of sleep, your body can rebuild and produce the testosterone you need for stronger erections.

Step 10:

Get outdoors. Guys need the vitamin D from the sunlight for at least 20 minutes on the hands, face, and feet. Vitamin D is necessary for testosterone production. Make sure as well that vitamin D is in your diet from foods such as fish, cheese, egg yolks, and certain types of mushrooms.

Appendix G

Karezza in Four Easy Steps

With thanks to 'Reuniting – Healing with Sexual Relationships' and L. Kevin Johnson

STEP ONE: So once you've educated yourself and made a firm decision that you want to learn Karezza and move away from masturbation and fertilization-driven mating sex, the first step is to limit ejaculation to no more than twice a month. After six months or so you'll find that you have less and less desire to unconsciously deplete your life-essence by continuously emitting semen, especially when you have no intention of impregnating a woman.

You may eventually discover that regularly masturbating and ejaculating are not that important anymore. Then it becomes your choice whether or not you want to continue. But twice a month is probably the safest upward limit if you wish to keep doing it and remain healthy, youthful, and vibrant. Many Karezza men report that the desire for orgasm and ejaculation completely goes away after a few months and is no longer an issue.

STEP TWO: For now, decide to put off masturbation and sexual intercourse for at least a couple of weeks to give your brain chemistry a chance to settle down and re-stabilize (read "The Passion Cycle" at www.reuniting.info, which will explain scientifically why this is so important). Meanwhile, you can recruit your wife or find a girlfriend to help you with this critical transition phase of moving from mating to bonding sex.

Allow her to offer you a genital/penis massage at least 3 or 4 times a week during the next two weeks. Doing it every day is

181

okay too, but you really should willingly do something non-sexual for her in exchange, such as dinner and a movie out, affectionate (non-sexual) snuggling, back massage, foot rub, house work, etc. It should be something of her choice that would please her.

A genital massage session should last at least 20 minutes but not more than 45 minutes. The point of this is to help acclimate you toward receiving direct genital touching without getting "heated up" or aroused to the point that you want to encourage the urge to ejaculate. Permit the woman to keep her clothes on. Lie on your back, open your legs and relax. Have her apply some almond oil to her hand and let her gently and very, very slowly massage your scrotum, testicles, penis and perineum. Breathe slowly and deeply while she softly and tenderly pulls the skin of the scrotum and pubic hair. These light touches require that you remain still. Have her push slightly (with short fingernails if possible) into your groin at different places around your penis to release built up tension. Don't encourage her to stroke the penis! She can do gentle, light squeezing and releasing along the shaft and head.

Due to the fact that the male genitals have experienced a constant build-up of tension through orgasm and ejaculation, this type of gentle massage from a female greatly relieves soreness and pain in that area. It is very soothing and relaxing and releases oxytocin in the brain, which makes you feel bonded to your partner

If you are prone to getting heated up easily, then have a bowl of ice and a cold damp washcloth next to the bed. As soon as you feel that familiar horny, full feeling, which means the semen is beginning to load in your prostate, have her stop the massage and place the cold rag on your testicles and the sensation will eventually subside. Then your lady can go back to the massage.

Remember, if abstaining from masturbating is causing you to get "blue balls," it isn't an indication that you need to ejaculate. It simply means that your body is adjusting to retaining and reabsorbing the semen into the surrounding groin tissue. To soothe the discomfort, apply the cold compress for a few minutes whenever the pain arises. It took only about a week for me to overcome the soreness when I finally quit masturbating. After that, my body adjusted and all the symptoms of "blue balls" went away once and for all.

The purpose of the penis massage is to enable you to learn how to focus your awareness on the present moment, develop heightened sensitivity, channel your sexual energy to the woman, and appreciate her touch. It is an excellent training method to prepare you for Karezza sex. It doesn't matter if you have an erection of not. Don't worry about that. Most probably, you will discover that you are numb or insensitive in this area from years of pursuing vaginal thrusting and hard masturbating. You have to relearn how to "feel" gentle sensations and welcome the pleasant nurturing of a woman's affection. When you can successfully get through two weeks of several penis massages without ejaculating and can remain calm and relaxed, you're ready to move on to the next phase.

STEP THREE: If you've gotten through at least fourteen days without ejaculating, you can now try peaceful Karezza intercourse. Start with a little bit of soft, unhurried, affectionate cuddling and relaxed kissing. Then after a few minutes, if you're not too heated up, try a round of partial insertion of the penis into the vagina, one to two inches only. Be sure to apply a generous amount of oil, such as unscented almond oil, on both of you.

Lie down naked on your right side, have the woman lie on her back, her left leg over your left hip, and ever so slowly glide the penis into the vagina. What's really incredible about doing it this way is that initially, it's not even necessary to have an erection. If you use enough oil, the two of you can practice soft penetration and easily "pop it in." In some cases, the penis will then slowly expand and grow inside of her.

From this point on, you have to focus on relaxing and staying in "calm waters," that is, keep yourself from getting swept away by the temptation of moving and rubbing. It is best to remain perfectly still. Remember, avoid getting heated up. Though it may seem like this kind of sex is boring and pointless, if you stay with it and wait, you will learn that there is an incredible gift for both of you.

It is as if the genitals know what to do and all you have to do is relax and let it happen. Your job is simply to monitor whether or not you're feeling an eruption coming, a point in which you feel tempted to move closer to orgasm. If this starts to happen, pull out and go back to cuddling and lying still together. Use the cold cloth method if necessary. Make soft eye contact and slowly kiss each other. Allow the feelings in your body to settle down and then try again. Keep going until you can make it last

184

at least 30 minutes. When you can master this form of loving, it is possible to stay connected for even an hour or more!

Note: If you feel the semen is about to spew and it's inevitable that you are going to ejaculate, try this: pull out immediately; press down hard on your perineum (the soft hollow tissue between the anus and scrotum) with the tips of your third & fourth fingers. Hold the pressure with your fingers, push the tip of your tongue to the roof of your month, and breathe slow and deep in through your nose and out your mouth.

If you do emit some semen, you will significantly reduce the amount lost by employing this method. Should this happen, it would be wise to discontinue the love session until another time, because your prostate has now become loaded with semen.

My best advice here is learning not to rush things. Never seek to stimulate the woman or make her feel horny or aroused. Avoid oral sex as well as clitoral stimulation. In Karezza, women should also avoid orgasm (read "What If She Were Always In the Mood" to understand how peak orgasm for both men and women can cause separation in intimate relationships). It is better to be affectionate, attentive, kind, and loving.

Your goal in Karezza is to get an energy circuit of the male and female life-force energy riding between the two of you. It's not about stimulating the genitals so you can have a release. The objective is to send the sexual energy back and forth between you, not discharge it. That is the delight to be discovered in Karezza. That's when the deep connection begins to happen, which is the valuable treasure and gift that this form of bonding offers.

For now, it isn't even important to penetrate deep into the woman's vagina. What you want to do is stay relaxed; open, still, and partially inside of her for an hour or so. If you want to change positions, do it slowly. I would strongly suggest avoiding lying on top of her, such as in missionary position. For most men this will only trigger the ejaculation/mating sex urge. If the woman lies on top of you, that is often perfectly fine, because it enables you to completely relax and not have to hold yourself up. Side to side works well, too. Any position is okay as long as you can relax, stay comfortable and avoid any kind of tension.

The two of you may want to benefit from this kind of love-making four or five times a week, but remember that it takes time and patience to get to the level where you, as a couple, can feel the flowing circuit. There is a pleasant energy that radiates from your perineum (base chakra) into the woman's vagina, up toward her breasts, then out from her to your chest, down your spine back toward your genitals, then out into her again.

186

The first time this phenomenon happened to me, I was stunned at how profoundly nourishing it was. I then realized that in all the years of pursuing orgasms for me and my woman, I had missed this incredible miracle of consciously linking with another human in love and kindness.

STEP FOUR: At this stage in the process you may find that it is easy to move toward deeper penetration, which the ancients called "the garden of Love." The penis, no matter what length, creates an energetic connection with the cervix. It is not necessary for the head of the penis to make physical contact with the cervix. It is the energy exchange during deep penetration that begins to generate the profound feelings of intuitive connection between a man and a woman.

It may take several months before the two of you can achieve sustained, deep penetration, especially if she experiences pain in her vagina due to past sexual trauma or emotional insecurity. The point is to make slow and steady progress toward deeper levels of relaxation and awareness. Never shove your penis into her. Avoid thrusting in and out of her just to stimulate yourself. True male authority means possessing a calm, loving penis that is used as an energetic "sending" instrument, not as a desensitized "getting" device.

Over time you will discover that the penis has an innate intelligence. It knows what it is doing and will do special things at different times. It may gently swell and then quickly shrink in size depending on the energy that is present at the time. Sometimes it will not swell into an erection at all, while other times it will be huge and hard. It is at these times that you need to be extra sensitive to the woman and only inch it in slowly, then stop and let it rest in one spot.

Amazingly, the penis will probe, explore, and pulsate on its own. Your job at all times is to focus on your penis and use your

187

awareness to move the universal life-force energy you feel into her body. You must become a giver of this life energy. It is no longer about you getting pleasure for yourself, yet it is a truly pleasant experience. Open your heart and cultivate loving feelings and kindness to the woman. Karezza is actually a process of discovery toward the higher goal of achieving unity between a man and a woman. Do you want to bond with your woman or do you want to fertilize her?

In Conclusion

This understanding of the goal of Karezza sex as a way to bond with another human being took me quite some time to uncover and learn. Now I feel that this simple act of consciously joining the genitals together has the potential of achieving the most profound effect, in allowing us to fulfill our true function. Human beings are here to bring love into the world, to make an impact upon the consciousness of the society and the planet. Through harmonious, sacred bonding of man and woman, the spiritual seeds of a new understanding can begin to grow and expand, reaching out to touch the lives of all people everywhere. Because all Mind is One, I believe this way of lovemaking is a powerful and significant transformative force.

"In simplest terms, Karezza is affectionate, sensual intercourse without the goal of climax," Robinson says. "Intercourse is generally frequent, although not necessarily daily. But couples will typically also engage in daily "bonding behaviours" – non-erotic skin-to-skin contact, gentle stroking, and so forth."

"Removing the goal of orgasm puts the focus on sex as a sensual experience and puts couples in the moment, so they are thinking about giving and receiving pleasure, not just aiming to get to the end," body and soul sexologist Dr. Gabrielle

188

Morrissey says. "Research shows that when it comes to sex, people value the connection with a partner more than the physical release. Karezza, and practices like it, can shift that focus to the connection instead of couples constantly chasing the orgasm."

In order to have multiple orgasms, without ejaculation or, in other words, in order to open the way to erotic ecstasy, you should practice sexual continence during lovemaking. This means that you must develop both your capacity for control and erotic sensitivity.

The techniques that will be presented here, in the "Men" section, will help you avoid ejaculatory orgasms, which decrease and eventually exhaust one's "reserve" of sexual energy, and any chance of experiencing multiple orgasms.

These methods are efficient according to the perseverance with which you will practice them. There are men endowed with a greater erotic energy and sensibility, as well as with a greater capacity for mental control. But still, anyone can succeed.

Such men will be able to experience multiple non-ejaculatory orgasms after a few weeks of practice. Other men, whose erotic sensibility and mental control are less powerful, will probably need to practice for months or years in order to get to this stage.

Consequently, remember that you may become a master only through practice.

BREATH CONTROLIn order to control your sexual energy and to practice sexual continence, you have to breathe as deeply and as relaxed as you can. All martial arts and yoga lessons indicate that breathing is the key needed to control the body.

Breathing is generally an involuntary act, but it may just as well become a conscious act. In other words, we usually breathe without thinking about it, and also without changing the natural rhythm of our breath. If we did that, if we made our breath more profound, we could influence the cardiac rhythm.

For instance, after we run, we breathe rapidly and superficially, and consequently our cardiac rhythm reaches high levels. If we breathe slowly, the cardiac rhythm decreases.

To the extent to which lovemaking is concerned, a high rate of the cardiac rhythm usually indicates that you approach ejaculation. The conclusion comes naturally: the first step in controlling your ejaculation is controlling your breath.

ABDOMINAL BREATHINGMost of us breathe superficially, usually at the level of the thorax and clavicles (clavicular breathing), which leads to a poor oxygenation of the lungs.

For instance, newborn babies breathe abdominally. If you watch a baby sleeping, you will notice that his belly moves with each breath he takes. Abdominal breathing fills our lungs with air and allows us to replace the residual air, stagnating inside our lungs with fresh air.

This is the healthiest way to breathe, but most of us lost this ability because of stress and anxiety. Our way of breathing is limited to the upper part of the chest, and therefore we are said to breathe only at the level of the thorax and clavicles.

When we are happy and we laugh, we breathe abdominally. The following exercise will show you how to breathe abdominally, as you did when you were quite young.

When you practice the techniques suggested here, it is required that you breathe in through your nose, as the air is filtered and

190

warmed up at the passage through the nose. Remember that whenever you breathe in through your mouth, you inhale unfiltered and cold air.

TECHNIQUE

1. Sit on a chair, spine straight, feet on the floor, and head up.

2. Place your hands on your navel and relax your shoulders.

3. Inhale through the nose and feel how the lower part of the abdomen becomes distended with air, so that the navel is pushed forth. The diaphragm goes down.

4. Relax the chest while exhaling forcefully through the mouth so that the lower part of your abdomen is pulled back, as if you wanted to push the navel towards the spine. You will also feel how your penis and testicles are somewhat drawn upwards.

5. Repeat steps 3 and 4, 21 times.

A few minutes of daily abdominal breathing will teach your body to breathe deeply again in a natural manner, even when you sleep. While making love, this capacity is essential in order to prevent ejaculation and to expand the erotic sensations throughout your whole body.

Once you no longer ejaculate, it is important that you carry on with your breathing exercises, as they will help you make the sexual energy circulate through your body and sublimate it into volitional, affectionate, mental, and spiritual energy.

The abdominal breathing massages the internal organs and the prostate, and it relieves the sensation of pressure most men experience for the first time when they do not ejaculate.

191

This sensation of pressure of tension in the genital area, which appears to all beginners in the practice of sexual continence, is caused by the stagnation of the sexual energy in the pelvic area. If the energy is not sublimated, it will lead to irritability, confusion, and edginess.

Therefore, we wish to make it clear that the mere transmutation of the sexual potential energy (the transformation of sperm into sexual energy, or in other words, the retention of the sperm inside the body and its transformation into other substances) is not enough to experience multiple all-body orgasms.

This is only the first step. The next step is the sublimation of this sexual energy, or in other words, the actual flow of this energy through the subtle channel corresponding upwards to the spine (you may even have unusual perceptions through the spine during this process).

Therefore, abdominal breathing is a highly important technique, because it sets in motion the sexual energy and guides it through the spine, resulting in its sublimation from the genital area and in the suppression of the above-mentioned states of irritability, confusion, etc.

Another method to determine this flow of the sexual energy is the practice of Hatha Yoga postures, which also have the effect of eliminating the energetic nodes and facilitating the circulation of the energy through the whole body.

ABDOMINAL LAUGHTERIf you have had trouble with abdominal breathing (as most Westerners do), you may practice abdominal laughter as well.

What is abdominal laughter? It is the kind of laughter that makes your abdomen shake. It is genuine laughter shared with

192

your close friends. It is the laughter that makes you say your stomach hurts from laughing. This pain is due to the fact that we do not use these muscles very often.

TECHNIQUE

Sit comfortably on a chair, spine straight and feet on the floor.

Place your hands on the abdomen, and remember all the funniest moments of your life. When laughter begins, let it grasp your entire being until you feel your stomach vibrating.

This kind of laughter is extremely helpful, as it relaxes the diaphragm, makes you breathe abdominally, and generates a great quantity of energy that you may use later on.

Appendix H

Tantric sex exercises

Tantric sex is not necessarily about "doing it," it is about building up the sexual tension, that super-horny feeling that you sometimes get and can do nothing about. The idea is to make foreplay the focus, keeping your mind off the orgasm and focusing on the sensations around you. Those sensations are touch, feel, scent, sounds, movements, temperatures, and more. The effort by the participants will be avoiding penetration until you can no longer take not doing it.

These exercises are done with a partner:

Step 1: Set the tone with the lights and temperature of your favorite area to be together intimately. Turn the lights down, phones off, and music on. Make sure your stomachs are comfortably full. Your clothes can be on or off.

Step 2: Loosen your body from the stress of the day or stress of your life. Shake it off by shaking your arms and legs – get them a little tingly.

Steps 3: Get comfortable standing, lying down, or sitting up.

Standing up: This intimacy-building exercise is called the "heart breath," designed to tune you into each other. Stand opposite one another and look into each other's eyes, placing your left hand on your partner's heart. The partner should then place his/her hand over your left hand, and you should try to match each other's breathing for at least two minutes.

If sitting down, there are three possible sitting positions: 1) knee to knee and heart to heart; 2) sitting with one partner's legs wrapped around the other's torso; 3) one partner sitting in the other's lap with your legs wrapped around each other. With these positions, try to create as much skin contact as possible. This can also be done by wrapping your arms around each other.

Step 4: Just touch each other gently, but firmly, with one finger all over the body. Masturbation is allowed, but not to the point of orgasm. Doing body massages is great as well.

Step 5: Just stay in that position, breathe deeply, stay calm, and get aroused.

Step 6: Prime the charge by prolonging foreplay. Priming foreplay activates the nervous system to hold a greater charge.

Step 7: When the time is right, and you will know it, start getting it on. The trick here is to breathe slowly while doing it. As you ramp up to orgasm, make sure you continue to breathe slowly. Breathing slowly by both partners delays the orgasms, but when it happens, it will be longer and more in intense.

Appendix I
Grounding exercises

This is a standing barefoot exercise that you can do in the morning while brushing your teeth. While standing still, take a deep breath and feel your big toes. Push them into the floor. Release. Do the same with each of the remaining toes (first the pointer toe, then middle toe, then ring toe, then baby toe). Now press the outside of the foot into the floor. Then do the same with the outside of the heel, back of the heel, inside round of the heel (along the arch if you can), and back to the big toe.

With all the driving and sitting, that society has created for us, this exercise will get you feeling your feet again, which is important for balance, stability, reconnection to your lower body, and overall wellness.

This is a great exercise to calm you down and allows you to let go of the day's stresses. Combine that with conscious breathing and this standing exercise works wonders for stress.

Appendix J

Whole body exercises for sex stamina and increasing your testosterone

Remember guys, long-endurance sports lower T. Short bursts of high intensity interval training are most likely best. Increasing levels of T are associated with short, high-intensity workouts, no longer than 60 minutes.

A beginner exercise to start boosting T might be one set of 5 squats, followed by 5 pushups, 5 sit ups, 4 squats, 4 pushups, 4 situps, 3 squats, 3 pushups, 3 sit ups, 2 squats, 2 pushups, 2 situps, 1 squat, 1 pushup and 1 sit up.

This will take probably 5 minutes. Certainly you can find 5 minutes in a day to start making a difference in your life.

You can gradually add one level to this beginner training every week, until you are ready for the following:

Sprints – 5-10 sprints for 6-10 seconds.

Bounding – jumping up stairs or onto a solid box surface.

Pull-ups.

Walking lunges.

Mountain climbers.

Spiderman pushups.

Better results come with full body multi-joint exercises.

Sample of a daily menu:

Breakfast:

Mancakes, 2 servings - 2 organic eggs with ½c cottage chees and & ½ c gf oats with cinnamon and vanilla ½ tsp each. Mixed in blender

 1 serving of almonds and 1 serving of organic blueberries.

Mid morning snack:

4 oz. of hummus with rice crackers or carrot sticks.

Lunch:

4 oz of salmon on 1-cup of arugula, with lemon/oil dressing.

Mid afternoon snack:

One avocado, split in half, drizzled with lime juice; spoon it out.

Dinner:

One baked chicken breast medium, skinless, with steamed broccoli, asparagus, and carrots.

Dessert is frozen raspberries in a blender with a tablespoon of honey.

Appendix K

Testicular Self-Exam

- If possible, stand in front of a mirror. Check for any swelling on the scrotal skin. It is best to do this when the testicles are warm and descended, as in after a shower or bath.
- Examine each testicle with both hands. Place the index and middle fingers under the testicle with the thumbs placed on top. Firm but gently roll the testicle between the thumbs and fingers to feel for any irregularities on the surface or texture of the testicle.
- Find the epididymis, a soft rope-like structure on the back of the testicle. If you are familiar with this structure, you won't mistake it for a suspicious lump.
- Thank you to the Testicular Cancer Society

Appendix L

Anger - dealing with it

There are many things that can be done to relieve anger. You could beat a punching bag, buy a hundred glass plates or cups from the dollar store and throw them at your wall, have psychotherapy, receive acupuncture, use Heilkunst homeopathy, do a liver detox, or do the following:

Determine what is it specifically that angers you? Identify clearly what it is that set you off. It can't be "Henry pisses me off!" What is it specifically about what Henry does or doesn't do that pisses you off? Write a list with each thing you perceive him to have done or not done "to you" to piss you off, with each thing being 2-4 words long.

Now this is where you have to work through a little deeper. To understand the next part of this you have to understand the idea of the laws of the universe. There are 12 laws of the universe that govern how the universe works. The one that we want to use with this section is the law of polarity – that we do not live in a one-sided world and that there is an opposite always present. Examples of this law would be that just as there is a sunrise, there is a simultaneous sunset; where there is happy, there is sad; with a positive on a magnet there is a negative. No one trait ever exists without its complimentary opposite being present as well. It is true with every trait.

What is the opposite of angry to you? When you identify your word/trait that is the opposite of angry (let's say your word is "calm"), start with this: When you feel angry, where in your body do you have the sense of calm? Identify that area. Now go to a memory when you were calm. Got that memory? Okay. Where were you, and whom were you with? Now let's go to another memory when you were calm. Got that memory? Okay. Where were you, and whom were you with? Repeat this until the total amount of calm is equal to the perceived level of anger you are experiencing.

It is a very effective way of letting go of your anger in a safe and productive manner.

Appendix M

With thanks to Peter Kanaris from Kanaris Psychological Services, from which his material on Masters and Johnson's Senate Focus was used in this section:

These exercises were originally developed by Masters and Johnson to help couples experiencing sexual problems, but can be used for a variety of intimacy issues and to heighten awareness with any couple...

In the first stage the couple has two sessions or "dates" in which they take turns touching each other's bodies, but the breasts and genitals are off limits. The purpose of the touch is not as a turn-on or to be sexual, but to establish awareness of sensations by noticing textures, temperatures, and contours. The goal is simply to be aware of sensations of being touched by a partner. The person being touched is told to focus on what interests them and not on any guesses as to what their partner likes or does not like. The partner receiving the touching is instructed to focus on the sensation and on what feels more or less pleasing. The couple is instructed that if sexual arousal occurs, they are not to proceed to intercourse. Masters and Johnson recommended that the initial sessions of Sensate Focus be as silent as possible because talking can distract from the awareness of physical sensations. The partner being touched must let his or her partner know either verbally or non-verbally if any touching is uncomfortable.

In the second stage of Sensate Focus, touching is expanded to include the breasts and the genitals. The "active partner" is instructed to begin with general body touching and not to immediately move to the breasts or genitals. Once again, emphasis is on the physical sensations and not the expectation of the sexual response. Intercourse remains a prohibition. Orgasm is not the goal. The couple is asked to take turns during a "hand riding" technique as a means of non-verbal communication. By placing one hand on top of the partner's hand while being touched, one can indicate if he or she would like more or less pressure, a faster or slower pace, or a change to a different spot. Masters and Johnson cautioned that these non-verbal messages could be conveyed in such a way that the person being touched does not take full control, but adds some additional input to the touching, which is still done based upon the interests of the active person.

In the third phase of Sensate Focus, instead of taking turns touching each other, the couple is asked to experiment with mutual touching. The purpose of this exercise is to practice a more natural, real form of physical interaction (people often do not take turns touching and being touched and to help each partner shift attention to a portion of his or her partner's body and away from his or her own response. Couples are reminded that no matter how sexually aroused they feel, intercourse is still off limits.

The next stages of Sensate Focus are to continue with the mutual touching and then at some point move into the female on top position without insertion of the penis into the vagina. From this position the female can rub the penis with her clitoral region, vulva, and vaginal opening, regardless of whether or not there is an erection. In a subsequent session she may put the tip of the penis into the vagina if there is an erection, while focusing on the physical sensations and stopping and moving back to non-genital touching if either partner becomes orgasm oriented or anxious. After completing a session or two at this level, couples are often comfortable to proceed to intercourse without difficulty....

Most importantly, Sensate Focus is used as a springboard to successful intimate communication. It allows for deconstruction of unhealthy behaviours. It is an opportunity to reconstruct a functioning and intimate sexual relationship. It is the context in which successful sex therapy unfolds. Paradoxically, the de-emphasizing of the sexual response seems to create conditions that actually facilitate it.

Appendix N

Breathing into your balls

This exercise is designed to ground you and have you get in touch with your sexual energy.

Always best with this exercise is to be in a quiet area, lit or dark (your choice). You can be sitting or lying down with your eyes opened or closed – just be in a comfortable position.

Once settled, and when breathing is relaxed and paced for two minutes, then begin.

When starting to take a deep breath, it is important to visualize energy coming in as a column of light between your eyes and above the bridge of the nose as the air comes in through both nostrils.

As your breath enters and joins the light, they travel a path together past the heart, collecting a stream of energy from the heart, through the stomach, across the small intestine and bladder, continuing on into the perineum, and then into the balls. One breath-in does this half of the pathway.

On the breath-out through the mouth, visualize the energy/light leaving your testicles and travelling outside of the body to join the space between your eyes, where it initially entered. It is there that the light collects until your breath is finished going out and you are ready to breath in.

Repeat this process for 5 to 10 cycles. When you are done, let your breath relax, breathe normally, slowly start to pay attention to the space around you, and go on with your day that much more grounded.

Check 'Em out

You know what to do
It's up to you to see it through
Talk to all your friends
Talk to all the men

Grab your balls
Check 'em out
That's what this is all about

Testicles are not immune
Cancer doesn't care about age or race
Time to talk about TSE
It's not taboo can't you see?

It's called 'self test'
So check 'em out
That's what this is all about

We need to take care of all our stuff
Silence has gone on long enough

It takes balls to check your bag
It takes you to take a stand
We all need to spread the word
Make a difference, you're a man

Grab your balls
Check 'em out
That's what this is all about

Grab your balls
Check 'em out
That's what this is all about

Grab your balls
Check 'em out
That's what this is all about

Words and music by David Shackleton 2014

www.ingramcontent.com/pod-product-compliance
Lightning Source LLC
Chambersburg PA
CBHW070353290526
45790CB00004B/1478